Cell growth and division

a practical approach

Edited by
Renato Baserga

Department of Pathology and
The Fells Research Institute,
Temple University Medical School,
Philadelphia, PA 19140, USA

IRL PRESS
—at—
OXFORD UNIVERSITY PRESS
Oxford New York Tokyo

IRL Press
Eynsham
Oxford
England

First Published 1989
Reprinted 1990

British Library Cataloguing in Publication Data

Cell growth and division.
  1. Animals. Cells. Development. Molecular
biology
  I. Baserga, R. (Renato) II. Series
  591.8′761

Library of Congress Cataloging in Publication Data

Cell growth and division.
  (Practical approach series)
  Includes bibliographies and index.
  1. Cell culture. 2. Cells—growth. 3. Cell division.
I. Baserga, Renato. II. Series. [DNLM: 1. Cell
Division. 2. Cells, Cultured. 3. Growth Substances.
QH 605 C392]
QH585.C46 1988 574′.07′24 88-31964

ISBN 0 19 963026 7 (hardbound)
ISBN 0 19 963027 5 (softbound)

Previously announced as:

ISBN 1 85221 114 8 (hardbound)
ISBN 1 85221 115 6 (softbound)

Printed by Information Press Ltd, Oxford, England.

cell
divi

a ᴐ ɔ ach

PRAC...H

Series editors:
Dr D Rickwood
Department of Biology, University of Essex
Wivenhoe Park, Colchester, Essex CO4 3SQ, UK
Dr B D Hames
Department of Biochemistry and Molecular Biology,
University of Leeds, Leeds LS2 9JT, UK

Affinity chromatography

Animal cell culture

Antibodies

Biochemical toxicology

Biological membranes

Carbohydrate analysis

Cell growth and division

Centrifugation (2nd Edition)

DNA cloning

Drosophila

Electron microscopy
in molecular biology

Gel electrophoresis of nucleic acids

Gel electrophoresis of proteins

Genome analysis

HPLC of small molecules

HPLC of macromolecules

Human cytogenetics

Human genetic diseases

Immobilised cells and enzymes

Iodinated density gradient media

Light microscopy in biology

Lymphocytes

Lymphokines and interferons

Mammalian development

Microcomputers in biology

Microcomputers in physiology

Mitochondria

Mutagenicity testing

Neurochemistry

Nucleic acid and
protein sequence analysis

Nucleic acid hybridisation

Oligonucleotide synthesis

Photosynthesis:
energy transduction

Plant cell culture

Plant molecular biology

Plasmids

Prostaglandins
and related substances

Protein function

Protein structure

Spectrophotometry
and spectrofluorimetry

Steroid hormones

Teratocarcinomas
and embryonic stem cells

Transcription and translation

Virology

Yeast

Preface

Cell growth means different things to different people. For many investigators, cell growth is just a question of growth factors, that is of the environment that surrounds the cells, in culture or *in vivo*. For others, cell growth is a problem of gene expression, that is of the genes and gene products that interact with and respond to the growth factors in the environment. For all of us, though, whether we purify growth factors or clone genes, the success of our work depends upon the assays we use; which takes me back (many years, unfortunately) to when I was a graduate student, and one of my professors told me that my experiments would only be as good as the assays used.

Cells constitute the basis for any assay of cell growth and (with few notable exceptions) the cells used are cells in culture. And there lies the rub, as cells in culture are fickle: *'qual piuma al vento'*, as Verdi would say. Thrown, somewhat brutally, into a hostile environment, cells in culture respond with a number of tricks to ensure their survival. Some (like human diploid fibroblasts) maintain a rigorous growth control but offer a stubborn resistance to transformation. At the opposite end of the spectrum, HeLa cells have jettisoned all growth controls and can be reduced to a state of no growth only by the drastic expedient of removing all proteins, a stage that closely resembles death. In between lies all kinds of cell lines, each of them with different growth requirements, different stabilities and different ranges of behaviour. Hence each cell line requires a different assay, and it would be foolish to expect that blood lymphocytes (the best G_0 cells on our planet) should behave in the same manner as HeLa cells. This book attempts to define these different assays in selected animal cell lines. I have tried to include some of the cell lines most frequently used as well as those that are less popular, concentrating on those that show growth regulation. The book should be useful to cell biologists, but particularly to molecular biologists who are interested in growth factors, growth-regulated genes and transformation.

I would like to thank all the contributors to this book, who actually sent their chapters *almost* within the deadline, and the staff of IRL Press, who have displayed an interest in the proceedings which is almost unique in publishers of scientific books.

Renato Baserga

Contributors

D. Barnes
Department of Biochemistry and Biophysics, Environmental Health Sciences Center, Oregon State University, Corvallis, OR 97331, USA

R. Baserga
Department of Pathology and The Fells Research Institute, Temple University Medical School, Philadelphia, PA 19140, USA

V. J. Cristofalo
The Wistar Institute, 3601 Spruce Street, Philadelphia, PA 19104, USA

E. Ernst
Department of Biochemistry and Biophysics, Environmental Health Sciences Center, Oregon State University, Corvallis, OR 97331, USA

D. Ewton
Biology Department, Syracuse University, Syracuse, NY 13244, USA

E. Ferris
Department of Cardiology, Children's Hospital Medical Center, Boston, MA 02115, USA

J. R. Florini
Biology Department, Syracuse University, Syracuse, NY 13244, USA

D. Greenwood
Department of Biological Sciences, University of Pittsburgh, Pittsburgh, PA 15260, USA

B. J. Lange
The Children's Hospital of Philadelphia, 34th and Civic Center Boulevard, Philadelphia, PA 19104, USA

D. Loo
Department of Biochemistry and Biophysics, Environmental Health Sciences Center, Oregon State University, Corvallis, OR 97331, USA

S. McGoogan
Department of Biological Sciences, University of Pittsburgh, Pittsburgh, PA 15260, USA

B. Nadal-Ginard
Department of Cardiology, Children's Hospital Medical Center, Boston, MA 02115, USA

P. D. Phillips
The Wistar Institute, 3601 Spruce Street, Philadelphia, PA 19104, USA

J. M. Pipas
Department of Biological Sciences, University of Pittsburgh, Pittsburgh, PA 15260, USA

M. B. Prystowsky
Department of Pathology and Laboratory Medicine, University of Pennsylvania, Medical Laboratory Building G3, Philadelphia, PA 19104, USA

C. Rawson
Department of Biochemistry and Biophysics, Environmental Health Sciences Center, Oregon State University, Corvallis, OR 97331, USA

J. Rheinwald
Division of Cell Growth and Regulation, Department of Cellular and Molecular Physiology, Harvard Medical School, 44 Binney Street, Boston, MA 02115, USA

M. J. Smyth
Cell Biology Group, Life Sciencies Division, Los Alamos National Laboratory, Los Alamos, NM 87545, USA

S. Shirahata
Department of Biochemistry and Biophysics, Environmental Health Sciences Center, Oregon State University, Corvallis, OR 97331, USA

A. Srinivasan
Department of Biological Sciences, University of Pittsburgh, Pittsburgh, PA 15260, USA

G. L. Stein
Department of Cell Biology, University of Massachusetts Medical Center, Worcester, MA 01655, USA

J. L. Stein
Department of Cell Biology, Univeristy of Massachusetts Medical Center, Worcester, MA 01655, USA

W. Wharton
Cell Biology Group, Life Sciences Division, Los Alamos National Laboratory, Los Alamos, NM 87545, USA

Contents

Abbreviations

ALL	acute lymphoblastic leukaemia
ANLL	acute non-lymphoblastic leukaemia
ATCC	American Tissue Culture Collection
BCGF	B-cell growth factor
BSA	bovine serum albumin
CEE	chick embryo extract
CK	creatine kinase
CPDL	cumulative population doubling level
CSA	colony-stimulating activity
CSF	colony-stimulating factor
DEX	dexamethasone
DMEM	Dulbecco's modified Eagle's medium
DMSO	dimethylsulphoxide
EBV	Epstein–Barr virus
EGF	epidermal growth factor
FCS	foetal calf serum
FGF	fibroblast growth factor
FITC	fluorescein isothiocyanate
GCT	giant cell tumour
GM	granulocyte–macrophage
HDL	high density lipoprotein
HS	horse serum
HTLV	human T-cell leukaemia virus
IL	interleukin
INS	insulin
LCM	lymphocyte-conditioned medium
β-ME	β-mercaptoethanol
MEM	minimum essential medium
MLR	mixed lymphocyte reaction
MSA	multiplication stimulating activity
PAI	plasminogen activator inhibitor
PBS	phosphate-buffered saline
PDGF	platelet-derived growth factor
PHA	phytohaemagglutinin
PPLO	pleuropneumonia-like organism
PPP	platelet-poor plasma
SFME	serum-free mouse embryo
STI	soybean trypsin inhibitor
TCA	trichloroacetic acid
THR	thrombin
TPA	tetradecanoylphorbol-13-acetate
TRS	transferrin

CHAPTER 1

Measuring parameters of growth

RENATO BASERGA

1. INTRODUCTION

This chapter deals with the various parameters of growth and how they can be measured.

A tissue can grow by: (i) increasing the number of cells; (ii) increasing the size of the cells; or (iii) increasing the amount of intercellular substance. Since the intercellular substance of a tissue is usually a secreted product of the cell, for example collagen, it can be considered, so to speak, as an extracellular extension of the cytoplasm. We can therefore consider an increase in intercellular substance as a variation of an increase in cell size and thereby reduce tissue growth to two mechanisms, growth in size and growth in the number of cells. This is true regardless of whether we are dealing with normal or abnormal growth. However, although both mechanisms may be operative, increase in cell number is by far the most important component in either normal or abnormal growth. Cells in culture can also grow either by increasing their size or by increasing their number.

There are also static and dynamic ways of measuring growth and cell division. For instance, counting the number of cells in a Petri dish tells us how much that cell population has grown. It does not tell us whether or not the cells are still proliferating. Other methodologies (autoradiography with [^{3}H]thymidine, flow cytophotometry, etc.) are necessary if we wish to examine cell proliferation and its pertubations in more detail.

In the following sections I will give a few simple techniques for measuring cell growth, and I will try to stress their interpretation and their limitations.

2. GROWTH PARAMETERS

From the foregoing, it is clear that there are several ways of measuring parameters of growth. The question is: which parameter of growth does one wish to measure? Take, for instance, a typical experiment in which one wishes to determine the effect that a growth factor has on a population of cells in culture. It is often stated in seminars and papers that a certain growth factor is mitogenic, but the only evidence we are shown to document its mitogenicity is a labelling index (with [^{3}H]thymidine) or, even worse, incorporation of radioactive thymidine into acid-soluble material. Mitogenic means that it induces mitosis: that is that cells divide and increase in number. Incorporation of [^{3}H]thymidine measures DNA synthesis, not cell division. The two processes often go together, but they can also

1

be separated (for a review, see ref. 1). If we wish to determine the effect of growth factors (or of any environmental change) on cell proliferation, that is their ability to stimulate or inhibit cell division, the best method is very simple, and it is to count the number of cells before and after treatment, possibly at 24-h intervals.

2.1 **Counting the number of cells in cultures**

(i) Prepare the dishes in which the cells have been grown (this example is for 100-mm dishes).

(ii) Prepare a trypsin solution, either 0.25 or 0.1% (see note i below) in Hanks' balanced salt solution (containing no Ca^{2+} or Mg^{2+}).

(iii) Pour Hanks' solution (no Ca^{2+} or Mg^{2+}) into 50-ml tubes.

(iv) Remove the medium from the dishes and set aside.

(v) Wash with 10 ml of Hanks'; remove.

(vi) Add 10 ml of trypsin solution and leave for 30 sec–1.5 min (note i).

(vii) Remove the trypsin solution and let stand at room temperature for 2–3 min (note i).

(viii) Add the growth medium (which includes 5% calf serum), 10 ml. At this point the cells will detach from the surface. In these days of very expensive serum we use the conditioned medium obtained from step (iv) to inhibit trypsin instead of fresh growth medium.

(ix) Mix well using a sterile pipette, drawing the cell suspension up and down the pipette 5–10 times.

(x) Count in a haemocytometer (*Figure 1*), by depositing a few drops of cell suspension under the coverslip. Use the four corners to count cells, divide by 4 and multiply by 10^4 to obtain cells/ml.

For instance, if you count 140 cells in the four corners:

$$140/4 = 35; \ 35 \times 10^4 = 3.5 \times 10^5 \text{ cells/ml}$$

Notes

(i) Trypsin strength varies from one batch to another (regardless of what the manufacturer says) and, in addition, sensitivity to trypsin is different in different cell lines. Therefore, it is impossible to give a single optimal trypsin concentration. One has to go by trial and error, and the same comments apply to steps (vi) and (vii). The goal is to obtain a suspension of single cells with as little as possible cellular debris.

(ii) The amounts can be appropriately scaled down if one uses smaller Petri dishes. The amounts of growth medium we use *to grow* cells (not to trypsinize them) are 20, 8 and 3 ml respectively, for 100-, 60- and 35-mm dishes. In short-term experiments (24–48 h), the amounts of growth medium can be reduced to half the indicated volumes, resulting in considerable savings.

(iii) Counting the number of cells in a solid tissue is somewhat more complicated. The difficulty here is to obtain a satisfactory suspension of single cells. Perhaps, for solid tissues, DNA amount (see Section 2.3) is the best available method.

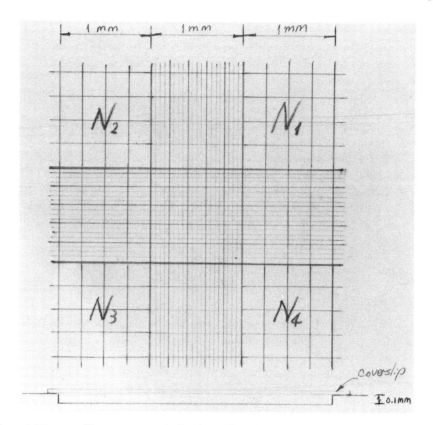

Figure 1. Diagram of haemocytometer's chambers. The grid is what one sees under the microscope. Below the grid is a cross-section of the haemocytometer, indicating that the space between the grid and the coverslip is 0.1 mm. Each N square thus has a volume of 0.1 mm^3. Multiplying the average number of cells per N square by 10^4 gives the number of cells per ml. Further corrections are necessary if the cell suspension has been diluted before counting.

2.2 Indirect measurements of cell number

The number of cells per dish is logically the best parameter to use to indicate whether or not a population of cells is growing. There are, however, alternatives. For instance, one can measure the amount of DNA per dish, or the amount of RNA, or the amount of proteins, or count the number of mitoses.

2.3 DNA amount

The amount of DNA per dish can be determined, for instance, at 24-h intervals after plating. It is also a very simple procedure; indeed, for someone trained in biochemistry, it is easier than counting the number of cells. Since the amount of DNA per cell is *usually* constant (in mammalian cells, 6×10^{-12} g/diploid cell in G_1), the amount of DNA per dish is an indirect measure of the number of cells. The G_1 amount of DNA in somatic cells is generally referred to as the $2n$ amount.

Cells in S-phase or in G_2 have increased amounts of DNA with respect to G_1 cells, but, if a population is truly growing, the increase in total amount of DNA per dish will go way beyond the error caused by the individual variations due to the distribution of cells throughout the cell cycle. DNA amount is probably the method of choice for solid tissues. Using the amount of DNA per cell given above, one can calculate (from the amount of DNA per μg) the number of cells per μg of tissue. A rough (very rough) estimate is that 1 μg of tissue contains 5×10^8 cells, but this estimate will vary greatly with the type of tissue.

A possible source of error is polyploidy, that is an increase in the amount of DNA per cell from $2n$ to $4n$ or even $8n$, in which case one could have an increase in the amount of DNA per dish without a concomitant increase in the number of cells. This is not a frequent occurrence, but it can happen, for instance, in certain pathological conditions, in response to certain drugs or when a large proportion of cells is blocked in G_2 (reviewed in ref. 1).

There are several methods for determining the amount of DNA in a culture dish or in a tissue. I prefer the classic method of Burton (2).

2.4 RNA amount

For a given cell type, the amount of RNA ought to be constant. As with DNA, G_2 cells will have roughly twice the amount of RNA as G_1 cells. As a measure of cell number, however, RNA amount is less accurate than DNA amount because, in several conditions, cells can grow in size and increase their RNA amount without cell division (3). Indeed, RNA amount is a good indicator of cell size, rather than cell number. Most of the cellular RNA that is measured by bulk chemical methods or individual cell histochemical methods is ribosomal RNA (rRNA ~85% of total cellular RNA). Since rRNA forms a part of the ribosome, on which protein synthesis is carried out, it seems logical that RNA amounts ought to be a reasonable indicator of cell size, a hypothesis that has been empirically confirmed. Classic methods for the determination of RNA amounts can be found elsewhere. If one wishes to determine the amount of RNA in individual cells, one needs expensive equipment that, in addition, require technical expertise to operate. I prefer computerized microspectrophotometry (4) to flow cytofluorimetry, but both are complex and unless one is a devoted cell biologist who loves to look at single cells, my advice is that when these instruments are needed one should seek collaboration. Flow cytometry has already been discussed in a previous book from this series (5).

2.5 Protein amount

The same comments apply here as to RNA amount. The amount of cellular protein is a reasonable indicator of cell size, but a poor indicator of cell number.

3. MITOSES

Surely, if one is looking at the mitogenic effect of a substance, the number of mitoses ought to be the best indicator of such an effect. However, mitoses are fleeting; in most cells they last only 45 min and, unless one looks at precisely the right moment, one may miss them. Furthermore, the duration of mitosis can increase in certain cells, especially in transformed cells (6). Everything else being equal, if the duration of mitosis in cell line A is twice that of cell line B, the number of mitoses in A will also be twice that of B, although the two cell lines may grow at the same rates. In tissues, the number of mitoses per 1000 cells (the mitotic index) is a reasonable measure of the *proliferating activity* of a cell population, but not of its growth. For instance, in the crypts of the lining epithelium of the small intestine, there are many mitoses. Fortunately for us, the small intestine in the adult individual does not grow, because, for every new cell produced in the crypts, one dies at the tips of the villi. So, in any given cell population, one must distinguish between cell division and increase in cell number.

There are also technical problems. To begin with, if we wish to determine the mitotic index of cells in culture a 22-mm^2 coverslip will have to be placed into the culture dish (see below for the preparation of coverslips). Mitoses can then be counted directly on the coverslips after fixation and staining (see below). Staining and counting of mitoses directly on plastic surfaces is not advisable. However, suppose that a wave of mitoses occurse 25–27 h after stimulation of a quiescent cell population with growth factors. One may miss it, unless samples are taken practically every hour. More economic in terms of time and money is to add a drug that will arrrest cells in mitosis. One can then count the percentage of cells that *accumulate* in mitosis over a certain period of time. The three main drugs for this purpose are colcemid, colchicine and nocodazole.

Colcemid and colchicine are very similar but the latter is more toxic. For mitotic arrest, the optimal concentration of colcemid is 0.16 μg/ml for human cells or 0.04–0.08 μg/ml for rodent cells. I like to leave the drug in for 4 h, then fix the coverslips. By dividing the time period into 4-h blocks (for instance, 16–20 h; 20–24 h; 24–28 h) after serum-stimulation, once should be able to get a pretty good idea of the mitotic activity of a cell population. If cells are left in colcemid (and especially colchicine) for more than 4 h, cell damage occurs with loss of mitotic figures.

Nocodazole (7) offers the advantage that it can be used for longer periods of time, 16–24 h. We use it at concentrations of 0.04–0.2 μg/ml, and mitoses, clearly identifiable, continue to accumulate. Depending on the cell line, and up to 12–16 h, nocodazole arrest is reversible (so is colcemid-arrest but only up to 4 h).

3.1 To clean coverslips for tissue cultures

(i) Pour chromic sulphuric acid (enough to cover the coverslips) into a large Petri dish.
(ii) One by one, place each coverslip in the acid. Leave to soak for 30–45 min.
(iii) Remove the acid and place the coverslips in a beaker. Let water run over them for 1–2 days.
(iv) After 1–2 days rinsing, rinse again with deionized water.
(v) Then take three large Petri dishes. Fill the first with methanol; the other two with 80% ethanol. Place all the coverslips in methanol; then individually

rinse each coverslip — first in one dish of ethanol, then in the other. Lay them out to dry on paper towels or wipe with gauze pads. Then autoclave. All handling after the chromic sulphuric acid is done with tweezers.

(vi) Coverslips are placed in tissue culture and, at the desired times, they are removed, washed three times in buffer and then fixed in methanol at $-20°C$ for 15 min. The coverslips are then mounted (cells up!) on a regular glass slide for convenient handling, using ordinary nail polish.

3.2 Cells arrested in mitosis by nocodazole

(i) Coat the coverslips (four per 100-mm Petri dish) with poly-L-lysine (Sigma 3000 mol. wt) at 1 mg/ml dissolved in Hanks' (calcium, magnesium free solution). Leave for 24 h.

(ii) Remove the polylysine solution and allow the coverslips to dry.

(iii) Plate 5×10^5 cells per 100-mm dish in normal growth medium, each dish containing two polylysine-coated coverslips.

(iv) After 18 h remove the medium and add fresh growth medium plus 0.1–0.2 μg/ml of nocodazole dissolved in dimethylsulphoxide (DMSO).

(v) Leave for the desired period of time and then fix using the method given above and stain with Giemsa/Sorenson's buffer.

4. DNA SYNTHESIS

It is often desirable to measure DNA synthesis instead of cell proliferation. This is especially true if one wishes to study G_1 events leading to the replication of DNA or if one wishes to know the fraction of proliferating cells in a given cell population. Measurements of DNA synthesis are also much more impressive than counting cell number. For instance, a mitogenic stimulus may double the number of cells in a Petri dish in 24 h. In the same time, the fraction of cells labelled by [³H]thymidine will go from 0.1 to 90%, virtually eliminating the need of statistical analysis.

The method of choice here is high-resolution autoradiography with [³H]thymidine. I will first outline the technique and then discuss its advantages and disadvantages.

4.1 Autoradiography (modified from Baserga and Malamud, 8)

Coverslips for autoradiography are prepared as outlined in Section 3.1. Cells are grown on coverslips and fixed as described in Section 3.1.

4.1.1 *Preparing tissue sections for autoradiography*

(i) Mounting the section on precleaned slides rubbed just prior to use with fresh egg albumin (egg white). (Avoid commercial albumin since it contains phenol, a reducing agent which causes a high background.)

(ii) Cut tissue sections of 3–10 (usually 5) μm.

(iii) Deparaffinize slides; set up 11 staining dishes:
  (a–d) xylene — 5 min each.
  (e) 50% xylene and 50% absolute ethanol — 5 min.
  (f–j) 100% alcohol — 3–5 min, then 95, 70, 50, 35%, always for 3–5 min.
  (k) distilled water — until slides are ready to be dipped. Do not allow to stand in water longer than 0.5–1 h.

4.1.2 *Dipping technique*

(i) *Equipment.*
  Slides to be dipped carrying a coverslip or tissue section complete with numbers applied with Indian ink at least 1 h before dipping, plus two trial slides
  Slide boxes with bags of Drierite (Bakelite slide boxes, 25-slide capacity) Drierite, gauze sponges
  NTB2 (Nuclear Track Emulsion) (Eastman Kodak Co., Rochester, NY)
  Scotch Brand Pressure Sensitive Tape (Minnesota Mining & Manufacturing Co., St Paul, MN)
  X-ray film envelopes, or aluminium foil
  Red china marker
  L-shaped galvanized tray
  Timer
  Coplin jar half-filled with distilled water at 40°C
  Glass stirring rod
  Scissors
  Plastic bag
  Log (batch no., slide no., date dipped, date developed, date stained, type emulsion, etc.)
  Water bath at 40°C
  Darkroom — use only Kodak Safe-Light 'Wratten' Series red lamp with 15-W bulb, with red filter, only when necessary.

(ii) *Procedure.*
  (a) Melt the emulsion by placing it into constant temperature bath at 40°C for approximately 90 min.
  (b) Hydrate the tissue slides or mounted coverslips in distilled water not more than 0.5 h before dipping (see above).
  (c) If the emulsion is to be used undiluted, pour it into a beaker and keep in a water bath at 40°C.
  (d) If the emulsion is to be used diluted: in complete darkness, except for a safe-light, fill the rest of the Coplin jar (see Equipment), with emulsion. (This 1 : 1 dilution may be discarded at the end of the experiment.)
  (e) DO NOT LET ANY LIGHT FALL ON EMULSION (turn safe-light towards wall).
  (f) Dip in two clean trial slides to test the consistency of emulsion.
  (g) Dip the experimental slides back to back, for about 2 sec, vertically with

frosted ends up into emulsion. Separate the slides and place on an L-shaped tray, frosted ends forward and up. Be careful not to scrape the side of the emulsion container or touch the surface of slide; it may cause mechanical exposure.

(h) Slides should air-dry in 20–30 min, but may take longer in a small damp room. Test trial slides to see if they are dry.

(i) When the slides are completely dry place ten or less, in a bakelite box. Seal the closure edge with black tape. Wrap the box securely in light-tight film envelope or two layers of aluminium foil, then completely seal with black tape. Write the batch no. with a china marker on the tape. Place the boxes in a plastic bag, wrap it around them and fasten it closed.

(j) Place the batch in a refrigerator to allow for exposure.

4.1.3 *Exposure*

The exposure time of autoradiographs depends on the type and amount of isotope used. Mouse tissues treated with $10\,\mu Ci$ of $[^3H]Tdr$ per mouse require about 10–12 days' exposure time.

Cell cultures: $0.02\,\mu Ci/ml$ of $[^3H]$thymidine for 24 h labelling require 3–4 days' exposure. For short labelling pulses (30 min or so) use $0.5\,\mu Ci/ml$ and 3 days' exposure. If you are in a hurry, simply increase the concentration of $[^3H]$thymidine, but remember, long exposure of cells to high concentrations can cause radiation damage.

4.1.4 *Developing*

(i) *Equipment.*

Water bath at 18°C including two buckets of crushed ice and thermometer

Darkroom, using only a 15-W bulb with red filter

Six staining dishes placed in a water bath

Staining trays, timer

Distilled water

D-19 Developer, make up fresh every week, store in a brown bottle at room temperature, dilution 595 μg in 3.8 litres, or 156 μg in 1 litre. Always filter before use (Eastman Kodak Co., Rochester, NY)

F-10 Fixer, make every 3–4 weeks and store in a brown bottle at room temperature (97 μg/500 ml) (Eastman Kodak Co., Rochester, NY — 1 lb package — 3800 ml H_2O at room temperature), filter and dilute 1:1 with distilled water before use.

(ii) *Procedure:* darken room.

(a) Fill the staining dishes with changes solutions (see below).

(b) Place the slides in racks.

(c) Slightly dirty the changes solutions by running an empty tray through them (it sounds magical, but it works better).

(d) Change solutions every ten slides.

(e) Changes: (1) developer (*undiluted — 5 min*); (2) distilled water — brief rinse; (3) diluted fixer — 8 min; (4) distilled water — 5 min.

(f) Dry (air-dry); store in dust-proof boxes until ready for staining.

4.1.5 *Staining*

Here two techniques are given, one using haematoxylin-eosin and one using Giemsa, both of which are useful for cells in culture, smears or tissue sections.

(i) *Giesma staining.*
   - (a) Place well-rinsed autoradiographs in buffered distilled water (pH 6.8) for 1 h; leave to dry.
   - (b) Stain with Giemsa, pH 4.8, for 1 h (Giemsa stock solution, 2.5 ml; methanol, 3 ml; distilled water, 100 ml; 0.1 M citric acid, 11 ml; 0.2 M disodium phosphate, 6.0 ml).
   - (c) Rinse, air-dry and mount.

(ii) *Haematoxylin and eosin.*
   - (a) *Equipment*

     It is important that ALL reagents be filtered before use.
      - (1) Mayer's haematoxylin: 1 μg haematoxylin (mol. wt 356.34, Eastman Organic Chemicals) plus 0.2 μg $NaIO_3$ (mol. wt 197.901, Fisher Scientific Co.) plus 50 μg ammonium alum (aluminium ammonium sulphate, mol. wt 906.688, Fisher Scientific Co.), plus 1000 ml H_2O; will last about 2–3 months.
      - (2) Eosin Y: eosin Y (water- and alcohol-soluble, Fisher Scientific Co.) — stock solution: 5.0 g, plus 1000 ml of H_2O. For use dilute 1 ml of stock solution with 99 ml of H_2O.
      - (3) 1% sodium acetate: $NaC_2H_3O_2$ (Baker's Reagents, mol. wt 82.04) — 10 g, plus 1000 ml of H_2O.
      - (4) Water bath at 18°C (will need ice).
      - (5) Staining dishes and trays.
   - (b) *Procedure*
      - (1) Prepare an 18°C water bath with nine staining dishes. All reagents must be cooled to 18°C before procedure can begin.
      - (2) Never use more than five slides per tray.
      - (3) Remember to filter all reagents before use.
      - (4) Do not use the same haematoxylin for more than two trays (ten slides). Change everything after one tray.
      - (5) Every time a batch of autoradiographs is stained, the optimum time of staining is determined by use of a standard slide, which is stained first. The other slides can be stained only after this optimum time is determined.
      - (6) Changes: (i–iii) Soak autoradiographs in three changes of distilled water for 10 min each (total 30 min).
         (iv) Stain with haematoxylin for 10–12 min. Check the intensity of

stain under microscope. If not sufficient, stain and check again
after another 1 min. Usually the staining time with haematoxylin will be found to vary between 2 and 3 min.

   (v) Rinse in water.

  (vi) Blue in 1% sodium acetate for 5 min. Check under microscope. Repeat procedure if bluing is not satisfactory.

 (vii) Rinse in water.

(viii) Stain with eosin for 1 min. Check under microscope. Repeat if
necessary.

  (ix) Rinse in water and air-dry.

(7) Remove excess emulsion from back of slide with razor blade.

(8) OPTIONAL — if autoradiographs are to be mounted, place slides
in xylene, mount with Permount and coverglass.

Figure 2 shows an example of a completed, acceptable autoradiograph of cells
labelled with [³H]thymidine.

(iii) *Glossary*. Trial slides are blanks used only to check when the emulsion has
melted and when it is dry after dipping. A standard slide is a coverslip (or
smear) of cells that have been labelled with [³H]thymidine and *are known* to
give good autoradiographs. For instance, once a year, we label a batch of
HeLa cells with [³H]thymidine for 24 h; a sample is taken and autoradio-

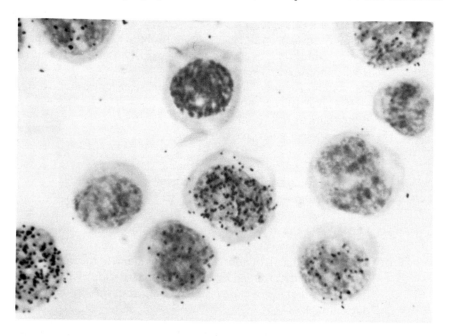

Figure 2. Autoradiograph of a suspension of tumour cells labelled with [³H]thymidine. The cells were labelled for 30 min; only cells in DNA synthesis during the labelling period show grains over the nucleus.

graphed. If the autoradiography is of good quality, we make many slides carrying these HeLa cells and place one slide in all batches of autoradiographs to be done. This standard slide will serve as a monitor to check that the autoradiography procedure has been carried out correctly. We find that the use of standard slides decreases the occurrence of frustration.

Pitfalls. Be careful of artifacts. Chemical or mechanical blackening of the emulsion can occur. The latter can be easily identified by its random distribution; the former can be more tricky. Formalin, for instance, if not carefully washed out, can give autoradiographic images that are quite convincing (but the cytoplasm will also be covered by grains). With tissue sections, airlocks can mimic an autoradiographic image.

5. DETERMINATION OF PARAMETERS OF GROWTH BY AUTORADIOGRAPHY WITH [^{3}H]THYMIDINE

The proliferating activity of a cell population can be estimated in a qualitative way by the thymidine index, that is by the percentage of cells labelled by pulse exposure to [^{3}H]thymidine. However, the thymidine index, like the mitotic index, gives only a rough estimate of the amount of cell proliferation in a tissue or in a cell population. For a more precise study, it is desirable to measure other parameters (refer to *Figure 3*). The three cell cycle parameters that are based on the use of

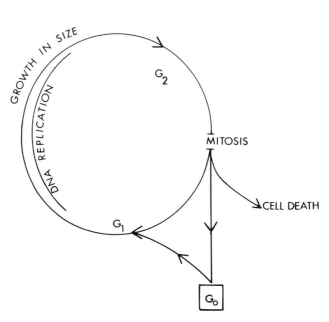

Figure 3. Diagram of the cell cycle. At completion of mitosis, cells have three options; (i) to differentiate and die without further division; (ii) to enter a non-proliferating G_0 state, from which they can re-enter the proliferative pool if an appropriate stimulus is applied; and (iii) to cycle again, through G_1, S-phase (DNA replication), G_2 and mitosis. While DNA replication occurs in a discrete period of the cell cycle, growth in size of the cell is a continuous process.

[³H]thymidine and autoradiography are:

  (i) the growth fraction;
 (ii) cell cycle analysis;
(iii) the time at which a block in the cell cycle is effective.

These three various ways of approaching the growth parameters of cells in culture, or in the living animal, will be described separately.

5.1 Growth fraction

The fraction of cells in a cell population that are actively participating in the proliferative process can be determined *in vivo* (9) and *in vitro* [(10) tissue cultures] by continuous labelling with [³H]thymidine.

5.1.1 *In vitro*

In exponentially growing cell populations the growth fraction is close to 100% and gives very little information. Thus, if we take almost any cell line that grows with a doubling time of 24 h or less during the exponential growth phase, and we label it with [³H]thymidine for 24 h, the number of labelled cells would be close to 100%. The odd unlabelled cells can probably be considered as dying cells.

The fraction of labelled cells becomes important in experiments dealing with the effect of growth factors on the state of quiescence of a cell population. For instance, many cell lines stop proliferating when they reach confluence. It is desirable to test the extent of quiescence in a cell population in culture by exposing the cells to [³H]thymidine. The interval usually taken, more for expediency than for any logical reason, is 24 h. Populations of cells that can be made quiescent should, at this point, have very low labelling indexes, that is not more than 1% of the cells ought to be labelled by a 24-h exposure to [³H]thymidine. At this point if one wishes to determine the effect of growth factors on the proliferation of this particular cell population, all one has to do is to add the growth factors together with [³H]thymidine and let the experiment run for 24 or 48 h. Cells entering S-phase will incorporate [³H]thymidine and the thymidine index will reach a maximum point beyond which it will not increase. Usually, under these conditions, 80–90% of the cells can be stimulated to enter S-phase by appropriate stimuli. There is always a refractory fraction of cells that do not enter S-phase and this may be due to the fact that these cells have left the proliferative pool permanently or, alternatively, that the density of the population is such that they cannot divide. In either case the experiment is very simple and it consists of comparing the fraction of cells labelled by [³H]thymidine over the same period of time, 24 or 48 h, in unstimulated cultures and in cultures stimulated with growth factors.

Obtaining the growth fraction in living animals is more complicated. A simple injection of [³H]thymidine is not sufficient with living animals because in the animal the injected [³H]thymidine is promptly broken down in the liver, and an injection of [³H]thymidine is practically equivalent to a 30–45 min pulse in a tissue culture. To obtain a reasonable growth fraction in mice or rats one can either use repeated injections of [³H]thymidine, usually at intervals of 4 h, or continuous infusion. Both methods are feasible when the cell cycle of the dividing fractions of cells is reasonably short and when the amount of radioactivity injected is not objectionable. The maximum number of cells labelled by continous exposure represents the growth fraction.

5.2 Quiescent cells and stationary cell populations

At this point it is important to distinguish quiescent populations of cells and stationary populations of cells. Stationary populations of cells are defined as populations of cells in which the number of cells does not increase. In ordinary circumstances this also means that the growth fraction would be very low and that very few cells would be labelled by [³H]thymidine over a period of 24 or 48 h, but this is not always true. There are certain cell lines, especially transformed cell lines, that reach a stationary phase, that is the number of cells increases no further, but if they are labelled for 24 h with [³H]thymidine, one may find that up to 50% of the cells are labelled. This is often a characteristic of cells that are transformed with certain DNA oncogenic viruses but in general it is a characteristic of transformed cells. Puzzlingly enough this mystery has never been solved. Some scientists believe that this is due to the fact that there is continous death of transformed cells in dense culture which are replaced by proliferating cells. Other scientists believe that the incorporation of [³H]thymidine into stationary populations of transformed cells simply reflects an increase in the amount of DNA per cell without cell division. Whatever the cause the investigator ought to be cognizant of the fact that in certain instances there is a huge discrepancy between the number of cells that are labelled by [³H]thymidine and the increase in the number of cells.

5.3 Phases of the cell cycle

To determine the duration of the various phases of the cell cycle the following experiment is given as an illustration.

(i) Grow replicate cultures of BALB/c-3T3 cells on coverslips in Dulbecco's minimum essential medium supplemented with 10% fetal calf serum, glutamine and the usual antibiotics.
(ii) 48 h after plating, when the cells are still growing exponentially, expose all cultures to [³H]thymidine at a final concentration of $0.2\,\mu\text{Ci/ml}$.
(iii) After 30 min wash all cultures twice with balanced salt solution and then re-incubate in a non-radioactive medium similar to the one described above.
(iv) At various intervals after removal of [³H]thymidine, terminate groups of two or three cultures, remove the medium and wash and fix the coverslips for autoradiography, as described above. For the following discussion, refer to *Figure 3*.

When cultures thus treated are examined no mitoses are found to be labelled immediately after removal of [³H]thymidine, while about 35–40% of interphase cells are labelled. The information we have obtained is that DNA synthesis occurs only during interphase and not during mitosis. For a variable period after removal of [^{3}H]thymidine, mitoses are still unlabelled, and since mitosis lasts for only about 45 min there must be a period before mitosis during which cells do not synthesize DNA. This period is called the G_2 period, or post-synthetic gap. The length of the G_2 period is given by the interval between the time of exposure to [^{3}H]thymidine and the time at which 50% of mitoses are labelled. For instance, if 50% of the mitoses are labelled 2.5 h after removal of [^{3}H]thymidine, one can say that the average duration of the G_2 period is 2.5 h.

As the time interval between removal of [^{3}H]thymidine and fixation of cells increases the percentage of labelled mitoses increases rapidly to 100%. These are the cells that were in DNA synthesis at the time [^{3}H]thymidine was added. Because of variations in cell cycle length among individual cells, it is not infrequent that the percentage of labelled mitoses does not quite reach 100%. The very shape of the curve of labelled mitoses (8), is an indication that cell cycle times vary among individual cells. The percentage of labelled mitoses remains near 100% for a time which is roughly equivalent to the duration of the phase during which DNA was synthesized and which we call the S-phase. It drops again and the percentage reaches a low point and, finally, the percentage of labelled mitoses increases again. This second wave of labelled mitoses represents cells that are entering mitosis for the second time since exposure to [^{3}H]thymidine. The determination of the various phases of the cell cycle is based conventionally on the 50% points. Therefore, the 50% points between the two ascending limbs of the curve of labelled mitoses gives the length of the cell cycle, T_c. The interval between the 50% points between the first ascending limb and the descending limb of the curve of percentage labelled mitoses gives the average duration of the S-phase. We have already said how to measure the G_2 phase and if we now add 45 min of mitosis the difference between the sum of S + G_2 + M and the total duration of the cell cycle gives the duration of the G_1 period which is the period between mitosis and S-phase.

The curve of percentage and labelled mitosis is useful if one wishes to determine the effect of growth factors on given phases of the cell cycle. Usually G_1 is the most variable period. For a given cell line S-phase, G_2 and mitosis are fairly constant though exceptions have been reported to occur.

5.4 Short method for growth fraction and cell cycle phases

A second method is continuous labelling in tissue cultures, or repeated injections of [^{3}H]thymidine in the living animal. Let us take as an illustration the tissue culture system.

(i) Add [^{3}H]thymidine, 0.02 μCi/ml to exponentially growing cells.
(ii) Fix the coverslips at various intervals after addition of [^{3}H]thymidine (please note that in this case the [^{3}H]thymidine is not removed).
(iii) Determine the fraction of labelled cells as usual.

The fraction of cells labelled within 30 min after exposure of the cells to [^{3}H]thymidine essentially gives the fraction of the cell cycle occupied by the

14

S-phase. If, for instance, the fraction of cells labelled was 40, we would have a simple equation:

$$40 = 100 \times T_s/T_c.$$

Under the conditions of cumulative labelling, all cells entering the DNA synthetic phase become labelled, and the time at which 100% of the cells are labelled corresponds to the sum of $G_2 + M + G_1$, that is $TG_2 + TM + TG_1 = T_c - T_s$. Let us say that the time of 100% labelling was 12 h. The only problem left is to solve a set of simultaneous equations, that is

$$40 \ T_c = 100 \times T_s$$

$$T_c = 12 + T_s$$

by solving, $T_s = 8$ h and $T_c = 20$ h. This method is very simple and though it does not separate the duration of G_1, G_2 and mitosis it does give simultaneously the growth fraction, the length of the cell cycle and the length of the S-phase.

5.5 Flow of cells through the cell cycle

At times it is desirable to determine whether there is a block or a delay in a particular phase of the cell cycle. This can be caused by a drug, by treatment with inhibitory factors or, for instance, in temperature-sensitive mutants, by a defect in one of the gene products that are necessary for cell cycle progression.

(i) A block in S-phase, due to direct inhibition of DNA synthesis, can be detected by exposing the cells at various times after the treatment, to a 30-min pulse of [^{3}H]thymidine. An inhibition of DNA synthesis will result in a quick decrease in the fraction of cells that can be labelled by the pulse exposure, usually within 30 min of treatment.

(ii) If, instead, the decrease in the fraction of cells that are labelled by pulse exposure to [^{3}H]thymidine is delayed, then one should suspect a block in G_1, or even later. For instance, suppose that a given drug acts at a point in G_1 which is roughly located 4 h before the S-phase. If that drug is given and the cells are then pulsed with [^{3}H]thymidine at various intervals after treatment, what one will observe is that the percentage of labelled cells will remain constant, as in controls, for 4 h. As the cells that were upstream of the block are now inhibited and cannot enter S-phase, the fraction of labelled cells will start decreasing as the cells that were in S-phase are exiting into G_2, but the cells that were located 4 h before the beginning of S-phase are not entering. By looking at the time required for the decrease in the fraction of labelled cells one can locate the block in the cell cycle. If the block is in mitosis one would of course see a marked increase in the number of mitotic cells all of which remain unlabelled. Similar experiments can be devised to determine whether there is a block in the flow of cells from S to G_2, or from G_2 to M.

6. QUIESCENCE OF CELLS

A few words on this subject are in order as there is a lot of confusion about the meaning of quiescence, the definition of G_0 etc.

Whether one likes it or not there is a physiological state of the cells in which the cells do not go through the cell cycle, yet are capable of doing so if an appropriate stimulus is given. The difference between G_0 and G_1 is now firmly established by the fact that in G_0 certain growth-regulated genes are not expressed. The ones that are commonly used are c-*fos* and c-*myc*, but several other genes have been described that are induced when G_0 cells are stimulated to proliferate (11). The fact that gene expression changes justifies a separation of G_0 from G_1.

It follows that simply labelling the cultures with [^{3}H]thymidine is not an indication of quiescence. It simply says that a population of cells cannot enter S-phase, but it does not really tell us whether those cells are in the G_0 phase in which certain growth-regulated genes are not expressed. A good illustration was reported recently by Ferrari *et al.* (12) with WI-38 cells, which can be made quiescent simply by contact inhibition. If one takes WI-38 human diploid fibroblasts and plates them in growth medium supplemented with 10% foetal calf serum, the cells grow to confluence. Upon reaching confluence, and without any change of medium, they become quiescent, at least as far as can be judged by labelling with [^{3}H]thymidine. Addition of serum stimulates the re-entry of WI-38 cells into the cell cycle. However, although by the 7th day after plating the labelling index is extremely low, Ferrari *et al.* found that certain growth-regulated genes, for instance c-*myc*, are still detectable in Northern blots. The cells are not truly in G_0 and this was demonstrated by the fact that they were still responsive to platelet-poor plasma which is a progression factor. However, if the cells were left confluent for a few more days, up to the 12th day after plating, c-*myc*, p53, ornithine decarboxylase, and other growth-regulated genes were no longer detectable and, at this point, the cells were no longer responsive to platelet-poor plasma, although they still responded to serum. These experiments clearly indicate that the labelling index is not a good criterion of quiescence and that to be sure that the cells are in G_0 one has to look at the expression of growth-regulated genes.

7. REFERENCES

1. Baserga, R. (1985) *The Biology of Cell Reproduction*. Harvard University Press, Cambridge, MA.
2. Burton, K. (1956) *Biochem. J.*, **62**, 315.
3. Baserga, R. (1984) *Exp. Cell Res.*, **151**, 1.
4. Mercer, W. E., Avignolo, C., Galanti, N., Rose, K. M., Hyland, J. K., Jacob, S. T. and Baserga, R. (1984) *Exp. Cell Res.*, **150**, 118.
5. Morasca, L. and Erba, E. (1986) *Animal Cell Culture: A Practical Approach*, Freshney, R. I. (ed.), IRL Press, Oxford.
6. Sisken, J. E., Bonner, S. V., Grasch, C. D., Powell, D. E. and Donaldson, E. S. (1985) *Cell Tiss. Kinet.*, **18**, 137.
7. Zieve, G. W., Turnbull, D., Mullins, J. M. and McIntosh. J. R. (1980) *Exp. Cell Res.*, **126**, 397.
8. Baserga, R. and Malamud, D. (1969) *Autoradiography*, Harper & Row, New York.
9. Mendelsohn, M. L. (1962) *J. Natl. Cancer Inst.*, **28**, 1015.
10. Stanners, C. P. and Till, J. E. (1960) *Biochim. Biophys. Acta*, **37**, 406.
11. Kaczmarek, L. (1986) *Lab. Invest.*, **54**, 365.
12. Ferrari, S., Calabretta, B., Battini, R., Cosenza, S. C. Owen, T. A., Soprano, K. J. and Baserga, R. (1988) *Exp. Cell Res.*, **174**, 25.

CHAPTER 2

Primary and multipassage culture of mouse embryo cells in serum-containing and serum-free media

DERYK LOO, CATHLEEN RAWSON, TED ERNST,
SANETAKA SHIRAHATA and DAVID BARNES

1. INTRODUCTION

Rodent embryo cells *in vitro* are the model of choice for many investigators studying animal cell growth and cell division. Embryonic cells are often used simply because these cells grow better under usual culture conditions than do cells from older organisms. Rodent cells are commonly used because one can obtain large amounts of mouse, rat or hamster embryos of precisely known age and genetic makeup, making it easier to replicate experiments. In this chapter we detail procedures for the primary and multipassage culture of cells from mouse embryos. We describe methods for both conventional cell culture in which a serum supplement is used as a source of growth stimulatory factors (1–4) and serum-free cell culture in which serum is replaced by specific combinations of peptide growth factors, binding proteins, nutrients, and attachment proteins (5, 6).

Properties of the cells derived by the two methods are quite different, and each has distinct advantages and disadvantages for various kinds of experiments examining regulation of cell proliferation.

2. PRIMARY AND PRECRISIS CULTURE

Mouse embryo cells cultured in conventional, serum-containing media grow for a limited period, rapidly losing proliferative potential and eventually undergoing growth crisis or senescence (1). In this section we describe methods for the initial (primary) culture of mouse embryo cells and maintenance of these cells *in vitro* during the precrisis period. We also describe methods for primary and multi-passage culture of mouse embryo cells in a serum-free medium that allows extended culture in the absence of crisis (6).

2.1 Primary culture of late stage mouse embryos

2.1.1 *Disaggregation*

Female mice that are 16–18 days pregnant are commonly used for primary mouse embryo cell cultures (1–4).

(i) Kill by CO_2 asphyxiation or cervical dislocation. Ether, phenobarbital or other anaesthesia may introduce complications due to the effects of these chemicals on the cells to be cultured.

(ii) Swab the abdomen with 70% ethanol. Remove embryos as soon as possible after death of the mother.

(iii) Surgically remove the embryos under a tissue culture hood with sterile instruments. Generally one or two pairs of sharp scissors and two pairs of forceps are sufficient to complete all manipulations.

(iv) Place embryos in a 100-mm culture dish.

(v) Remove membranes surrounding each embryo.

(vi) Wash the pooled embryos with 10 ml of phosphate-buffered saline (PBS) without calcium or magnesium.

(vii) Transfer the embryos to a fresh dish and mince with scissors. Embryonic tissue is quite soft, and it should be possible within a few minutes to obtain a fine mince that can be pipetted.

(viii) Add 10 ml of trypsin solution (0.25% crude trypsin with 1 mM EDTA in PBS without calcium or magnesium) to the plate. Enzymatic treatments used in the isolation of cells from adult tissues, such as collagenase or collagenase–dispase treatment, generally are not necessary.

(ix) Incubate the plate containing the mince at 37°C. This incubation should not be carried out in a CO_2 incubator, because PBS is not bicarbonate-buffered. An alternative approach is to transfer the fine mince to a 50-ml sterile test tube and carry out the trypsin incubation in a water bath or to use a bicarbonate-buffered medium without calcium or magnesium.

(x) Monitor the progress of disaggregation by microscopic examination. Obviously, this is most easily accomplished if the incubation is done in dishes. Pipette or agitate the suspension periodically.

(xi) Terminate the incubation when the major portion of the cells observed are single cells.

(xii) Allow the larger chunks of tissue to settle out for a few seconds in a centrifuge tube by gravity.

(xiii) Remove the supernatant trypsin solution containing the cell suspension.

(xiv) Centrifuge the cells from the trypsin solution in a bench top centrifuge.

(xv) Resuspend in culture medium (usually with 10% serum; see below). If a high yield of cells is desired, the larger chunks that settle from the suspension may be further trypsinized by repeating the procedures described above.

Do not extend the initial incubation for long periods in an attempt to obtain a homogeneous single cell suspension because lengthy incubations will lead to cell death. In addition to reducing yield, overtrypsinization causes other problems. Dead cells will release chromatin, and the protease activity of the trypsin solution will destroy the DNA-associated proteins, leading to hydration of the freed DNA and a noticeable increase in viscosity of the suspension. It is best simply to avert this problem by limiting trypsinization, but in unavoidable instances DNase may be added to digest the released material. Some crude trypsin solutions may contain sufficient contaminating DNase to prevent this problem.

If long-term storage of embryonic cells is desired, the cell suspensions derived from embryos may be frozen in liquid nitrogen in medium with 10% glycerol or dimethylsulphoxide (DMSO) and serum, just as one would freeze a suspension of established cultured cells. It is important that the cells be reasonably disaggregated for good viability upon thawing, since large clumps of cells do not freeze or thaw evenly, leading to cell death.

2.1.2 *Culture in serum-containing medium*

Carry out the following steps after resuspending cells in medium.

(i) Estimate cell number by haemocytometer count.
(ii) Plate cells at a density of $1-2 \times 10^5/cm^2$. This plating density is much higher than densities which should be used at later passage, but it is necessary at this point because the majority of cells in the initial suspension will not attach or grow in culture.
(iii) Change the medium 8–16 h after plating in order to remove debris and cells that will not attach. A significant amount of non-adherent red blood cells may be present in the initial plating, depending on how well the embryos were washed in the early steps.

The basal nutrient media commonly used for mouse embryo cell cultures is Dulbecco's modified Eagle's medium (DMEM) or minimum essential medium (often supplemented with non-essential amino acids). Usually a 10% calf or foetal calf serum supplement is used and antiobiotics are added (e.g. 200 U/ml penicillin and 200 μg/ml streptomycin). Basal nutrient media formulations specifically designed for mouse embryo cells are also available (7). Although some medium formulations were developed for use with a 10% CO_2 atmosphere, investigators more commonly use 5% CO_2, and the bicarbonate concentration should be adjusted accordingly (e.g. 1.2 g/litre for DMEM 5% CO_2). The cells in initial culture will represent several different cell types, and morphology may be quite heterogeneous (*Figure 1*). The appearance of the cultures becomes more homogeneous upon multiple passage.

2.1.3 *Culture in serum-free medium*

For the culture of mouse embryo cells in serum-free medium, carry out procedures for disaggregation of tissues as described above, except that after centrifuging the cells from the trypsin solution, resuspend the cells in 10 ml of serum-free medium containing 1 mg/ml soybean trypsin inhibitor and then recentrifuge and resuspend in fresh serum-free medium for counting. The soybean trypsin inhibitor stops the proteolytic activity of the trypsin. Inhibitors in serum carry out this function when one uses serum-containing medium. Cell suspensions isolated from embryos for serum-free work can be frozen in serum-free medium containing 10% DMSO or glycerol, although viability is not as good as that of cells frozen in serum-containing medium.

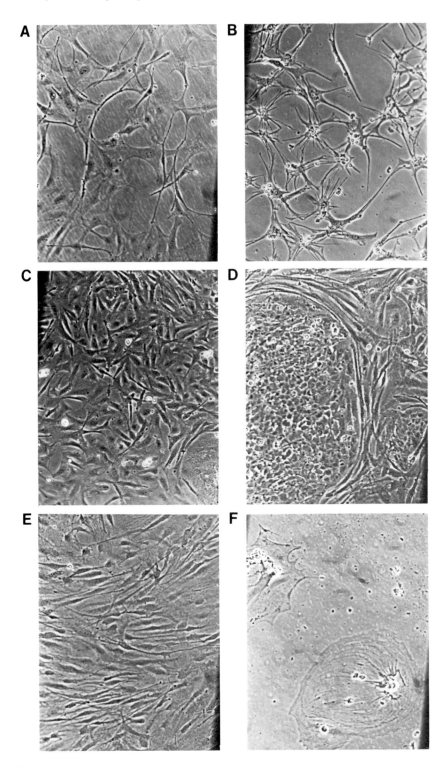

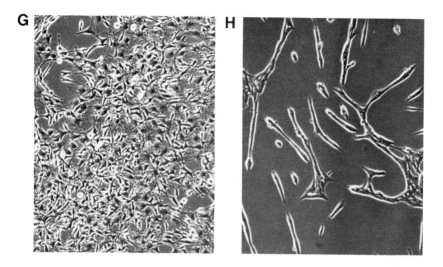

Figure 1. Photomicrographs of mouse embryo cell cultures. (**A**) and (**B**), primary culture at moderate density in serum-containing medium; (**C**) and (**D**), primary culture at high density in serum-containing medium; (**E**), third passage culture in serum-containing medium; (**F**), cells at crisis in serum-containing medium; (**G**), primary culture at high density in serum-free medium; (**H**), third passage culture at moderate density in serum-free medium.

The basal nutrient medium used for serum-free culture is a one-to-one mixture of DMEM containing 4.5 g/litre glucose and Ham's F12 (DMEM:F12) supplemented with 15 mM (Hepes), pH 7.4, 1.2 g/litre sodium bicarbonate, penicillin (200 U/ml), streptomycin (200 μg/ml) and ampicillin (25 μg/ml). Cells are routinely grown in a 5% CO_2–95% air atmosphere at 37°C. For the growth of cells under serum-free conditions, the culture dishes are pretreated with fibronectin and the basal nutrient medium is further supplemented with growth-stimulatory factors as described below.

For the preparation of concentrated stocks and the medium, purify water by triple glass distillation or passage through a Milli-Q (Millipore) water purification system immediately prior to use. Storage of water purified earlier for medium is not recommended, because even minimal microbial growth can lead to pyrogen contamination of the water, creating problems in serum-free culture. Serum components are capable of detoxifying functions, preventing the appearance of this problem in conventional culture.

Prepare F12:DMEM from powdered formulations (Gibco, Grand Island, N.Y.). The liquid medium may be stored for a maximum period of 3 weeks. Beyond this time degradation of some nutrients may lead to inconsistent results. Storage at −20°C is preferred. Store antibiotics and Hepes frozen as concentrated stocks (100×). Add these to the medium after dissolving the powdered formulation in a volume of water near that of the final volume.

Store all solutions in reusable plastic containers. Use polypropylene tubes for concentrated stocks and 250-ml polystyrene flasks for storing media. All pipettes and culture vessels should be disposable plastic. Toxic residual detergent remaining on glassware makes the washing and reuse of pipettes or glass containers risky for serum-free procedures. The tendency of some of the medium components to stick to glass surfaces also makes the use of glass disposable pipettes undesirable.

Culture the cells in the basal serum-free medium as described above supplemented with bovine insulin (10 μg/ml), human transferrin (25 μg/ml), human high density lipoprotein (HDL) (20 μg/ml), mouse epidermal growth factor (EGF) (50–100 ng/ml), and sodium selenite (10 nM) on human fibronectin-precoated dishes (20 μg/ml). Although platelet-derived growth factor (PDGF) was included in the original medium used in the derivation of some lines (6), PDGF effects are marginal in the presence of the other growth stimulatory supplements, and it is routinely omitted (although it is useful for serum-free growth of 3T3 cells). Do not add supplements directly to the medium and store for later use, as one might routinely do with a serum supplement. Insulin, fibronectin, HDL and other components do not remain active under such conditions. Instead, add insulin, transferrin, EGF and HDL directly to medium in individual plates or flasks as small aliquots from concentrated stocks immediately after plating cells.

Prepare insulin at 1 mg/ml in 20 mM HCl and transferrin as a 5 mg/ml stock in PBS or culture medium. Filter-sterilize both insulin and transferrin stocks after preparation. Insulin and transferrin can be obtained from Sigma, or a number of other sources. EGF and other peptide growth factors may be obtained as sterile, lyophilized powders from commercial sources (Bioproducts for Science, Indianapolis, IN; PDGF Inc., Boston, MA; R and D Systems, Minneapolis, MN; Collaborative Research Inc., Waltham, MA). Reconstitute with sterile water or buffered salt solutions as indicated by the vendors. Store sterile stock solutions of supplements in the refrigerator. Supplements may be stored long-term in the freezer in aliquots. Multiple freeze–thaws should be avoided.

HDL (density = 1.068–1.21 g/ml) is prepared by KBr ultracentrifugation separation (8).

(i) Start with freshly drawn citrated human plasma that has had platelets spun out. Approximately 1 unit of plasma (200 ml) is needed. It is important that the plasma be as fresh as possible and not frozen. Keep the plasma cold (4°C) until ready for procedure.

(ii) Adjust the density of plasma to 1.068 g/ml with KBr using the formula:

$$g\,KBr = \frac{ml(sample) \times (1.068 - 1.005)}{1.0 - 0.312}$$

Stir while adding KBr on ice for about 20 min or until the KBr has dissolved.

(iii) Spin in an ultracentrifuge at 105 000 g for 48 h, at 4°C, and decelerate slowly (70Ti rotor at 32 K r.p.m. in a Beckman ultracentrifuge). The tubes will look

stratified with a yellow clearly defined upper layer (LDL), a clear middle region and a brown/orange bottom region.
  (iv) Discard the upper layers and pool together the bottom portions of each tube by withdrawing the brown/orange fraction with an 18-gauge needle.
  (v) Adjust the density of this pooled plasma fraction to 1.21 g/ml with KBr using the formula:

$$g\,KBr = \frac{ml(sample) \times (1.21 - 1.068)}{1.0 - 0.312}$$

Add KBr and stir on ice as above.
  (vi) Spin at 173000 g (42000 r.p.m., in a 70Ti rotor) for 48 h, at 4°C, using slow deceleration mode. HDL is the intensely yellow band at the top of the tube.
  (vii) Pool the HDL from all the tubes and dialyse extensively against 0.15 M NaCl, 0.01% EDTA, pH 7.0–7.4.
 (viii) Filter through 0.45 μm membrane and assay for protein.

Yields should be approximately 20 ml of HDL with 15 mg/ml protein. Depending on the plasma source yields may be as low as 200 mg and as high as 400 mg. Addition of fatty-acid-free bovine serum albumin to 1 mg/ml is recommended; this may help prevent toxicity due to decomposition of HDL components. Divide the HDL into 1-ml aliquots and store at 4°C.

Fibronectin is prepared by gelatin-affinity chromatography (9). Gelatin–sepharose is available from Bio-Rad or Pharmacia. Bovine serum (Gibco), filtered (0.2 μm) before use, may be used as a source of fibronectin. Human or bovine plasma will provide higher yields. Bio-Rad gelatin affi-gel comes ready-to-use in PBS with 0.02% sodium azide as a preservative. The capacity for human plasma fibronectin is 1 mg/ml gel. The gel can be run at room temperature. Store all solutions at 4°C and warm to room temperature just before use. Column dimensions are 3 × 19 cm with 80 ml of affi-gel.

  (i) Wash the column with PBS and apply bovine serum. Up to twice the column volume of serum may be applied.
  (ii) Run serum into the column at 10–20 ml/h.
  (iii) Wash with PBS at 10 ml/h until OD_{280} is 0.05 or less.
  (iv) Wash with 1 M NaCl in PBS until the OD is again low.
  (v) Wash with one column volume of PBS.
  (vi) Prepare fresh 4 M urea in 0.05 M Tris, pH 7.5.
  (vii) Elute until the fibronectin peak is obtained, collecting 1-ml fractions. Store fractions on ice.
 (viii) Pool peak fractions and filter (0.2 μm).

Aliquots of 2 ml are stored at −86°C until required, then stored at 4°C while in use. To regenerate column, wash with PBS, 8 M urea in 0.05 M Tris, pH 7.5, PBS again, and store the column in PBS with 0.02% sodium azide at 4°C. Fibronectin is provided to cells by precoating flasks in the incubator with 4 ml of a solution of fibronectin at the indicated concentration in F12 : DMEM for 30 min and removal of the precoating solution prior to plating cells.

2.1.4 *Karyotype*

Mouse cells in primary culture are primarily diploid, although chromosomal abnormalities quickly develop upon multiple passage in serum-containing media, and some karyotypic aberrations can be detected, even in primary culture (1). The following karyotyping procedure, based on established protocols (10), may be applied to primary cultures (75-cm^2 flasks) or to later passages.

  (i) Change the medium the day after harvesting the cells. The cells should be in log-phase growth and 70–90% confluent.
  (ii) In order to induce metaphase arrest, remove the medium and replace it with fresh medium containing colcemid to a final concentration of 0.2 μg/ml. Colcemid stock solution is maintained at 10 μg/ml.
  (iii) Incubate the cells for 2.0 h at 37°C.
  (iv) Remove the medium and trypsinize cells.
  (v) Stop the proteolytic activity with soybean trypsin inhibitor and centrifuge cells from suspension.
  (vi) Remove the supernatant and resuspend the pellet by gentle agitation.
  (vii) Slowly and gently add 1 ml of 0.075 M KCl warmed to 37°C, followed by an additional 4 ml of this solution, added more quickly.
  (viii) Incubate for 20 min at 37°C.
  (ix) Slowly add 1 ml of fixative dropwise (3 : 1 methanol : glacial acetic acid, prepared fresh) with constant gentle agitation. This stops the swelling action of the hypotonic KCl.
  (x) Centrifuge the tubes at low speed.
  (xi) Remove most of the supernatant, leaving enough to resuspend the pellet.
  (xii) Slowly add 1 ml of fixative dropwise with constant gentle agitation to completely mix each drop.
  (xiii) Add 4 additional ml of fixative dropwise. At this point cells can be stored in a refrigerator for several days. Before continuing, stored cells should be warmed to room temperature for at least 15 min and centrifuged from suspension.
  (xiv) Remove and discard most of the supernatant, leaving enough to resuspend in the pellet.
  (xv) Add 1 ml of fixative dropwise with constant agitation. This can be added more quickly than in previous similar steps.
  (xvi) Add an additional 4 ml of fixative.
  (xvii) Let stand at room temperature for at least 15 min.
  (xviii) Centrifuge cells from suspension.
  (xix) Repeat the fixing procedure.
  (xx) Remove and discard as much supernatant as possible without disrupting the pellet.

To prepare the cell suspension for making slides, follow the following procedure.

(i) Add just enough fixative slowly, with gentle agitation, to obtain an optimal cell concentration. This statement is necessarily vague, since it takes practice to visualize what the optimal concentration will be. It is preferable to start with the concentration slightly too dense so that a few more drops of fixative can be added if needed after a test slide has been examined.

(ii) Dip a clean slide in distilled ice water; remove excess water by touching the slide to a paper towel.

(iii) Hold the slide at a 45° angle. Using a 9-inch Pasteur pipette, drop 3 drops of the cell suspension on the slide from a height of 6–18 inches. The drops should not overlap. The height can be varied depending on how well the metaphases spread on the slide.

(iv) Allow the slides to air-dry upright at an angle.

(v) Evaluate the slide using phase-contrast microscopy. Do not use oil immersion.

If the slide is too dense, dilute the cell suspension by adding a few drops of fixative. If the slide is too sparse, cells may be concentrated by centrifugation and resuspended. If the metaphases have many overlapping chromosomes, increase the dropping height or gently blow on the slide after adding the cells to enhance spreading. If most metaphases are not intact and chromosomes are scattered, decrease the dropping height. Once a satisfactory procedure and cell density on the slide has been achieved, make several slides and air-dry overnight.

Stain slides in Coplin jars containing 3% Giemsa (1.5 ml Giemsa stock solution in 48.5 ml of Sorensen's buffer, pH 6.8). Sorensen's buffer contains: 13.26 g of KH_2PO_4 and 15.2 g of $Na_2HPO_4 \cdot 7H_2O$ per 2 litres of water. Adjust pH with 1 M NaOH and store refrigerated. Giemsa stain should be made fresh each day.

(i) Stain slides for 7–10 min.

(ii) Dip slides in two rinses of distilled water and let them stand upright at an angle to dry.

(iii) Examine the slides microscopically using bright field. If chromosomes are too lightly stained increase staining time.

(iv) Once the slides are satisfactorily stained, allow to dry for at least 24 h.

(v) Add coverslips and score metaphases, using oil immersion lens.

2.2 Primary culture of earlier stage mouse embryos and rat embryos

Procedures for culture in serum-containing media of embryos as young as 12 days are identical to those described above. It is important to keep in mind that the amount of material obtained becomes progressively smaller with decreased embryonic age, and the properties of the cell types cultured may be different for embryos of different ages. Embryos younger than 12 days cannot be easily identified and removed without the help of a dissecting miscroscope. Similar approaches are applied for serum-free culture of earlier stage mouse embryos, but we have found the nature of the cells obtained from these sources to be quite different than the cells routinely cultured from 16-day-old embryos.

Procedures for culture of rat embryos in serum-containing media are essentially identical to those for mouse embryos. It should be noted, however, that 16- to 18-day-old rat embryos are at an earlier developmental state than are mouse embryos of the same age. We have not had reproducible success at serum-free culture of rat embryos by applying directly the procedures developed for mouse embryos. Presumably species-specific medium modifications are necessary.

2.3 Passaging and cloning mouse embryo cells

For routine passaging of cells in serum-containing media.

(i) Remove the medium and add a small volume of trypsin/EDTA solution.

(ii) Incubate at 37°C for about 5 min until the cells detach. Do not expect to remove all cells in the primary culture, because some cell types are so strongly adherent that it is unlikely they will be removed by simple procedures that will not harm the bulk of the cells. The percentage of cells that will detach upon routine trypsinization increases upon multiple passage, simply due to selection of removable cell types.

(iii) Add a volume of serum-containing medium equal to the volume of trypsin/ EDTA solution used and centrifuge cells from suspension.

(iv) Resuspend and replate in fresh medium.

The procedure for trypsinization of serum-free cultures requires some finesse. The serum-free derived cells are extremely sensitive to trypsin, and only a few seconds of exposure to the trypsin solution is necessary for removal from the flask. It is unnecessary to warm the trypsin solution to 37°C; room temperature is adequate. Cells should then be diluted into an equal volume of soybean trypsin inhibitor solution (1 mg/ml in serum-free F12:DME) and centrifuged from solution.

Only small volumes of trypsin and trypsin inhibitor are needed (e.g. 0.5 ml per 25-cm^2 flask). It is important to try to maintain high cell densities in the suspensions, because loss of cells occurs under serum-free conditions if dilute cell suspensions are manipulated. This is presumably due to adherence of the cells to pipettes and tubes. The problem can be reduced somewhat by the addition of 1 mg/ml bovine serum albumin to the trypsin inhibitor solution.

Cloning of primary or early passage cells in serum-containing medium is best accomplished with cloning rings, rather than by the limiting dilution method. Early passage cells do not tolerate culture at low cell densities. After removal with cloning rings, the cells should be placed in small wells (e.g. 24-well plates) in order to maximize cell density. Serum-free cloning is extremely tricky. Cloning rings are needed, but it is best to avoid the use of trypsin. Detach the cells by gentle pipetting over the cells in the ring of a solution of 1 mM EDTA in PBS without calcium or magnesium. Transfer the suspension of cells in this solution directly to a fibronectin-coated well with serum-free medium and the appropriate supplements. Do not attempt to centrifuge the cells from the suspension. Change the medium the next day.

3. CRISIS AND IMMORTALIZATION

Mouse embryo cells cultured for multiple passage in serum-containing medium undergo growth crisis (1–3). This is followed by degeneration of the culture or, in some cases, by the appearance of 'immortalized', genetically abnormal cells that are capable of indefinite growth in culture (*Figures 2* and *3*). Other changes observed upon immortalization of mouse embryo cultures include decrease in the percentage of methylated bases in the genome (11) and the expression of proliferin, a hormone-like factor initially identified as a serum-responsive protein in 3T3 cells (12). A very different pattern is observed upon multiple passage of mouse embryo cells in serum-free medium (6). Crisis is not observed and the cells that proliferate remain predominately diploid (*Figures 4* and *5*).

3.1 Extended culture in serum-containing media

The number of cell divisions before crisis in serum-containing media, the prominence of the crisis period and the nature of the immortalized cells that appear depends on the passaging protocol employed (*Figure 6*) (1). Cells maintained at low cell densities (e.g. $5 \times 10^3/cm^2$) enter crisis within 7–12 generations, and often the outgrowth of established lines or 'immortalization' is not observed.

Cells maintained at intermediate densities (e.g. $15 \times 10^3/cm^2$) undergo a well-defined crisis after about 20 generations, and generally give rise to established, non-tumorigenic cell lines after the crisis period. A passage regimen in which cells are plated at $15 \times 10^3/cm^2$ and passaged every 3 days approximates the protocol that gave rise to commonly-used 3T3 cell lines (1). For a given protocol (e.g. 3T3), strain differences exist regarding the number of divisions that will

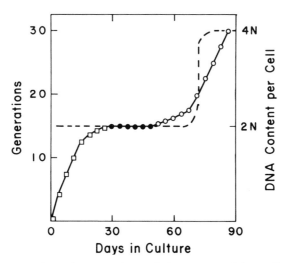

Figure 2. Multipassage culture of mouse embryo cells in serum-containing medium. Open squares indicate early growth phase; closed circles indicate crisis phase; open circles indicate growth of established cells from the senescent cultures; dashed line indicates DNA content per cell at each phase. See refs (1–4).

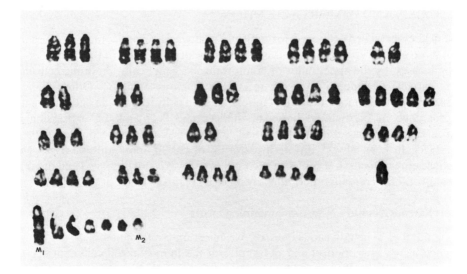

Figure 3. Giemsa-banded karyotype of typical established Swiss mouse embryo cell line derived by growth for 20 generations in serum-containing medium. This cell contains trisomies or greater numbers of several chromosomes, missing X or Y chromosome, small chromosomes of unidentifiable origin and large chromosomes derived from translocations. The chromosomes designated M1 and M2 are abnormal marker chromosomes present in all cells of this particular line.

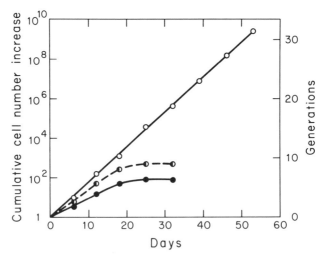

Figure 4. Growth of BALB/c mouse embryo cells upon successive transfer in serum-containing or serum-free media. Closed circles, cultures grown in medium supplemented with 10% calf serum; open circles, serum-free cultures grown in medium supplemented with insulin, transferrin, EGF, and HDL on flasks precoated with fibronectin; half-closed circles, cultures containing both the serum-free supplements and 10% calf serum. Cultures were initiated at 10^5 cells/25-cm^2 flask, medium was changed three or four days after plating and cells were passaged 6 or 7 days after plating. Cell number per flask was determined on the first day after plating and at passage and cumulative increase in cell number calculated from these determinations.

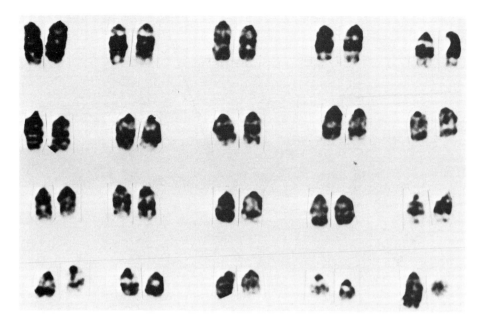

Figure 5. Giemsa-banded karyotype of typical Swiss mouse embryo cell line derived by growth for 200 generations in serum-free medium. This cell contains an apparently normal male karyotype.

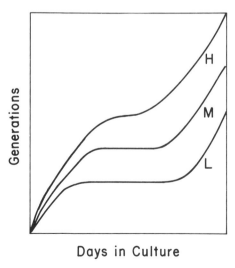

Figure 6. Representative comparison of growth of mouse embryo cells initiated and carried multi-passage in serum-containing medium at low (L), medium (M) and high (H) cell densities. See refs (1–4).

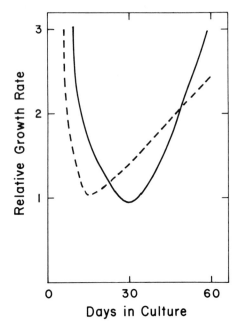

Figure 7. Representative comparison of multipassage growth of Swiss (solid line) and BALB/c (dashed line) mouse embryo cells initiated and carried multipassage in serum-containing medium in a 3T3 regimen. See refs (1–4).

occur in precrisis (1–4). For instance, BALB/c embryo cells in culture undergo fewer divisions and enter crisis earlier than Swiss embryo cells in culture under identical conditions (*Figure 7*).

Mouse embryo cells maintained at high densities (60×10^3/cm^2) may undergo as many as 30 divisions without measurable crisis (1). Such cultures may not undergo a visually perceptible crisis due to the high plating densities. These cultures give rise to established cell lines with regularity, but lines derived from cultures maintained under these conditions, such as the 3T12 cell line, are likely to be tumorigenic *in vivo* (13).

In almost all cases established mouse embryo cell lines derived in serum-containing medium are near tetraploid, generally with one or more marker chromosomal translocation and often with detectable deletions or minute chromosomes (*Figure 3*). A few lines exist in which karyotypic abnormalities are minimal, but these lines are still clearly aneuploid and may be genetically unstable (14, 15). In general, tumorigenic lines derived at higher plating densities will grow to higher saturation densities and require lower serum concentrations for optimal growth than non-tumorigenic established lines.

3.2 Extended culture in serum-free medium

Mouse embryo cells in primary culture in the serum-free medium (serum-free mouse embryo; SFME) proliferate without crisis (*Figures 1* and *4*). Similar results are obtained with both Swiss and BALB/c mouse embryos (6). Although the embryological relationship of SFME cells to serum-derived mouse embryo cell lines such as 3T3 is not clear, several lines of evidence suggest that SFME are mesodermally-derived cells, just as the serum-derived lines are thought to be. In addition to morphological appearance, SFME cells transformed with the *ras* oncogene (see below) and injected into athymic or inbred BALB/c mice produce undifferentiated sarcomas, based on histological examination of the tumours. SFME cells of both Swiss and BALB/c origin retain a predominantly diploid karyotype, and analysis of individual chromosomes by Giemsa banding techniques identifies no abnormal chromosomes (*Figure 5*).

The degree of variation of chromosome number around the modal diploid number of 40 is about the same for SFME cultures as for control primary cultures of cells in serum that are considered to be diploid standards ($\pm$ 2 chromosomes). Division infidelity of this kind is a consistent property of mouse cells *in vitro*. SFME cultures contain approximately 5% tetraploid cells; a percentage similar to that reported for precrisis mouse embryo cells maintained in serum-containing medium (1). When SFME cells are randomly cloned and the fastest growing colonies are picked, many clones contain translocations. This phenomenon is presumably due to selective pressures resulting from the cloning procedure employed, which requires cells to proliferate in plates seeded at very low cell densities.

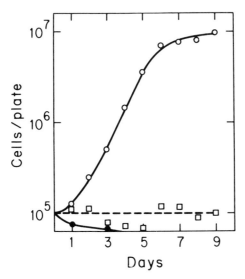

Figure 8. Effect of medium supplements on growth of serum-free derived mouse embryo cells. Open circles, serum-free medium with supplements as described in text; open squares, medium supplemented with 10% calf serum; closed circles, serum-free medium with all supplements except EGF. Swiss mouse embryo cells initiated and grown for approximately 200 generations in serum-free medium were used for the experiment. Cells were plated at 10^5/35-mm diameter dish. Cell number was determined at the indicated times after plating by counting cell suspensions in a Coulter particle counter.

Swiss SFME injected subcutaneously into 4- and 7-week-old athymic Swiss mice at 10^7 cells per mouse are non-tumorigenic. Similar results are obtained when BALB/c SFME are injected into inbred BALB/c mice. Oncogene-transformed or spontaneously transformed cells from serum-derived cultures (i.e. Kirsten murine sarcoma virus-transformed 3T3, 3T12) produce progressively growing tumours 1–3 weeks after injection of 10^6 cells per mouse. Omission of EGF from the serum-free culture medium results in cell death within a few days (*Figure 8*). Addition of serum to the culture medium leads to inhibition of cell growth, both in the presence and in the absence of the supplements used in the serum-free medium (*Figure 8*).

4. PLASMID TRANSFECTION AND SELECTION OF TRANSFORMED SFME CELLS IN SERUM-FREE MEDIUM

Calcium phosphate-mediated transfection of SFME with the human Ha-*ras* gene (pEJ6.6) or SV40 T antigen molecularly cloned into pBR322 (16, 17) allows the selection of cells capable of growing in serum-supplemented culture medium, either with or without additional supplementation with the factors used in the serum-free medium. These cells also grow well in the serum-free SFME medium in the absence of EGF and are tumorigenic in syngeneic or athymic mice. Frequency of stable transfection is approximately 3 in 10^6 cells.

No spontaneous, mock-transfected or control (mouse liver or pBR322) transfected colonies of SFME capable of progressive growth in serum-supplemented medium, either with or without additional supplementation with the growth-stimulatory factors used in the serum-free formulation, are observed. As controls for the study of oncogene-transformed SFME, we have transfected BALB/c and Swiss SFME with the pSV2neo plasmid, which contains a gene for a bacterial aminoglycoside phosphotransferase which inactivates the antibiotic G418 and confers on cells expressing this gene resistance to the toxic activity of the compound (18). G418-resistant SFME colonies isolated after transfection maintain parental phenotype with regard to tumorigenicity, serum inhibition and EGF requirement and maintain an otherwise normal, diploid karyology.

Below are detailed procedures for transfection and selection of oncogene-transfected SFME. Related procedures applicable to transfection with genomic DNA, selection and detection of oncogene-transformed serum-derived mouse embryo lines such as 3T3 and 10T1/2 may be found in Chapter 3.

4.1 Serum-free transfection procedures: calcium and strontium phosphate precipitates

These procedures are adapted from established approaches for introduction of exogenous DNA into cultured animal cells (19, 20). Solutions needed are a $2 \times CaCl_2$ (2.5 M calcium chloride in water) and $2 \times$ Hepes–PBS solution ($2 \times$ HBS), which contains the following in a total of 10 ml: 2 ml of 250 mM Hepes–NaOH at pH 7.2, 1 ml of 2.5 M NaCl, 1 ml of 18 mM Na_2HPO_4–NaH_2PO_4 at pH 7.2 and 6 ml of water. Store all stocks as sterile solutions. DNA for transfection should be in storage buffer (10 mM Tris, 5 mM NaCl, 0.1mM EDTA) at a concentration of 200 μg/ml to 2.0 mg/ml.

(i)   Plate cells in a 75-cm^2 flask at least 24 h prior to transfection.

(ii) Replace medium on cells with fresh medium 2 h prior to transfection.

(iii) Add 0.5 ml of 2 $\times$ HBS to a sterile polystyrene tube (Falcon 2058).

(iv) To a second tube add 50 μl of CaCl$_2$, 20–50μg of DNA in a maximum volume of 450 μl and sterile water to a final volume of 0.5 ml. Volumes of 10–200 μl of DNA solution work best. Shear the DNA sample to be used twice through a 25-gauge needle prior to use.

(v)   Add the DNA-containing solution drop by drop to the tube containing the 2 $\times$ HBS. To ensure gradual and gentle mixing, flick the tube by hand every one or two drops. A cloudy precipitate will form in the tube.

(vi) Allow the precipitate to sit undisturbed for 30 min at room temperature.

(vii) Shear the suspension twice through a 25-gauge needle attached to a tuberculin syringe to disperse the precipitate. The suspension should then appear cloudy.

(viii) Add the suspension dropwise to the flask containing cells with either the syringe or a Pasteur pipette. Tilt the flask and add the solution to the medium, rather than directly to the cells. Swirl the flask to ensure mixing of the precipitate before returning the flask to a horizonal position, in order to ensure that the precipitate falls evenly over the layer of cells.

(ix) Incubate the flask undisturbed on the bench top for 30–60 min.

(x)   Return the flask to the incubator and incubate for 3–5 h.

(xi) Change the medium after this incubation to fresh, complete serum-free medium.

Some cell types, including SFME, do not tolerate high concentrations of calcium well, and transfection efficiency or cell survival can be improved with the use of strontium phosphate as the DNA-containing precipitate (21). In this procedure the following modifications should be made. The 2 $\times$ HBS solution should be adjusted to pH 7.8 using NaOH and strontium chloride solution (2.5 M in water, pH 7.0) should be used in place of calcium chloride.

4.2 Selection of oncogene-transformed serum-free derived cells

Medium in the flask may be changed to apply the selective conditions 24–48 h after transfection. Alternatively, cells may be transferred to three to ten large dishes (100 mm in diameter) after transfection. In this case, plating cells on the dishes in the serum-free medium described, but with 5% calf serum added as well, improves efficiency of plating and colony formation when the selection is applied. Incubate the cells in serum-containing medium with supplements (including fibronectin precoat as described above) overnight before switching to serum-free medium in the absence of selective pressure. Following a 24–48 h incubation in the complete serum-free medium to allow cell proliferation in the dishes, remove this medium and replace with the selective medium. Selection for oncogene-transformed cells can be achieved by culture of cells in serum-free medium from which EGF is omitted or by culture of cells in medium containing 10% serum. Replace medium as needed, or at least every 7 days. When serum-free medium is changed infrequently, add insulin every 3 days.

5. CONCLUSION

The advantages of conventional mouse embryo culture in serum-containing media are well recognized. This approach is relatively simple and easy, as well as inexpensive. In addition, an established body of knowledge exists to make routine culture of these cells quite predictable. Even crisis, senescence and appearance of abnormal established lines, phenomena that one might consider to be disadvantages of conventional mouse embryo cell culture, are objects of major research efforts toward an understanding of the underlying biochemical mechanisms and the nature of the action of 'immortalizing' oncogenes.

Advantages of the serum-free approach described in this chapter include the derivation of lines that do not undergo crisis or exhibit abnormal karyotype and the considerable advantage of precise control of the components of the culture medium (5–7, 22–24). Prior to our work with mouse embryo cells, serum-free medium had also been shown to allow the extended growth of some other cell types without crisis (25, 26). In addition, the inhibitory response of these cells to serum represents a unique biological response not observed in conventional cultures.

Disadvantages of the serum-free approach include the relative expense of the components compared to serum-containing medium and the considerable time investment necessary in preparing medium and supplements and carrying out the tedious but essential steps in serum-free culture that are not required in conventional methods. For both serum-derived and serum-free derived lines, the embryonic origin of the cells cultured remains unclear, although in both types of culture it is likely that the cells are of mesodermal derivation.

6. ACKNOWLEDGEMENTS

We thank Kevin Nusser, Angela Helmrich, Constance Jackson, Paul Collodi and Yung-Jin Chung for help and advice and W. Peterson for karyotype analysis. Supported by NIH 40475, NIH 07506 and Council for Tobacco Research 1813. D. B. is the recipient of NIH Research Career Development Award CA01226.

7. REFERENCES

1. Todaro, G. J. and Green, H. J. (1963) *J. Cell Biol.* **17**, 299.
2. Aaronson, S. A. and Todaro, G. J. (1968) *J. Cell. Physiol.*, **72**, 141.
3. Jainchill, J. L., Aaronson, S. A. and Todaro, G. J. (1969) *J. Virol.*, **4**, 549.
4. Reznikoff, C. A., Brankow, D. W. and Heidelberger, C. (1973) *Cancer Res.* **33**, 3231.
5. Barnes, D. W. (1987) *Biotechniques*, **5**, 534.
6. Loo, D. T., Fuquay, J. I., Rawson, C. L. and Barnes, D. W. (1987) *Science,* **236**, 200.
7. Shipley, G. D. and Ham, R. G. (1981) *In Vitro*, **17**, 656.
8. Gospodarowicz, D. (1984) *Cell Culture Meth. Mol. Cell Biol.*, **1**, 69.
9. Ruoslahti, E., Hayman, E. G., Pierschbacher, M. and Engvall, E. (1982) *Methods in Enzymology*. Cunningham, L. and Frederiksen, O. (eds), **82**, 803.

10. Worton, R. G. and Duff. C. (1979) *Methods in Enzymology,* Jakoby, J. and Pastan, I. (eds), **58**, 322.
11. Wilson, V. L. (1983) *Science,* **220**, 1055.
12. Linzer, D. H. and Wilder, E. L. (1987) *Mol. Cell. Biol.,* **7**, 2080.
13. Aaronson, S. A. and Todaro, G. J. (1968) *Science,* **162**, 1024.
14. Farber, R. A. and Liskay, R. M., (1974) *Cytogenet. Cell Genet.,* **13**, 384.
15. Sasaki, M. S. and Kodama, S. (1987) *J. Cell. Physiol.,* **131**, 114.
16. Shih, S. and Weinberg, R. (1982) *Cell,* **29**, 161.
17. Pipas, J. M., Peden, K. W. C. and Nathans, D. (1983) *Science,* **290**, 1392.
18. Southern, P. J. and Berg, P. J. (1982) *Mol. App. Genet.,* **1**, 327.
19. Corsaro, C. M. and Pearson, M. L. (1981) *Somatic Cell Genet.,* **7**, 603.
20. Graham, F. L. and van der Eb, A. J. (1973) *Virology,* **52**, 456.
21. Brash, D. E., Reddel, R. R., Quanrud, M., Yang, K., Farrell, M. P. and Harris, C. C. (1987) *Mol. Cell. Biol.,* **7**, 2031.
22. Ham, R. G. (1984) *Cell Culture Meth. Mol. Cell Biol.,* **3**, 249.
23. Barnes, D. W. and Sato, G. H. (1980) *Cell,* **22**, 649.
24. Barnes, D. W. McKeehan, W. C. and Sato, G. H. (1987) *In Vitro Cell. Dev. Biol.,* **23**, 655.
25. Ambesi-Impiombato, F. S., Parks, L. A. and Coon, H. G. (1980) *Proc. Natl. Acad. Sci. USA,* **77**, 3455.
26. Brandi, M. L., Fitzpatrick, L. A., Coon, H. G. and Aurbach, G. D. (1986) *Proc. Natl. Acad. Sci. USA,* **83**, 1709.

CHAPTER 3

Effects of oncogene expression on cellular growth factor requirements: defined media for the culture of C3H 10T1/2 and NIH3T3 cell lines

DAVID GREENWOOD, ASHOK SRINIVASAN, SANDRA McGOOGAN
and JAMES M. PIPAS

1. INTRODUCTION

Established cell lines are usually cultured in a basal nutrient medium supplemented with serum. The serum provides the hormones, growth factors, nutrients and binding proteins that, as a whole, are conducive for the survival and proliferation of many cell types. However, serum is a complex mixture of diverse factors, many of which can have multiple effects, being growth stimulatory for some cell types and growth restrictive in others (1). Furthermore, these factors and the pathways through which they function, can act synergistically or antagonistically in a cell-type-dependent manner to influence the control of differentiation or proliferation. Recent work concerning the effects of growth factors on the proliferation and differentiation of specific cell types and on the dissection of signal transduction pathways involved in this regulation have made it increasingly necessary to achieve a more precise control of the culture environment.

Towards this end, several laboratories have developed serum-free media that support the proliferation of various cell types (for reviews see 2–4). This methodology, in which cells are cultured in a rich basal-nutrient medium supplemented with purified growth factors rather than serum, allows the investigator to control the concentration of individual factors. This is critically important for the culture of certain cell types that have unique factor requirements not provided by serum, or whose proliferation is inhibited by factors present in serum. The major technical difficulty arising from serum-free cell culture stems from the fact that each cell type has its own unique set of factor requirements and that different cell types may respond to the same factor in different ways. Thus, a culture medium must be tailored for each cell line being studied.

In our laboratory we are studying the effects of oncogene expression or activation on cellular growth factor requirements. Additionally, we have been interested in developing methods that allow the isolation of variant cell lines with altered responses to growth factors. For these experiments we have been using two established mouse embryo fibroblast cell lines, NIH3T3 and C3H 10T1/2. We have developed defined media that support the proliferation of these lines (1, 5, 6). In these formulations the cell lines will proliferate when plated at clonal density, divide at a rate slightly less than that in medium with serum, and reach the same saturation density obtained in serum-supplemented medium. Using such defined media it is possible to: (i) titrate the growth factor requirements of cell lines and compare the growth factor requirements of lines expressing various transforming genes; (ii) place selective pressure on cells by maintenance in media lacking required growth factors, enabling the selection of transformed cell lines which have altered responses to growth factors; (iii) assay for the presence of growth factor-like activity in conditioned media; and (iv) maintain certain primary lines for many passages without crises (7). This chapter will present methods dealing with the first two of these procedures.

2. GENERAL METHODS

2.1 **Laboratory facilities and equipment**

Cultures are maintained in standard CO_2 incubators at 37°C as usual. However, without the added buffering provided by serum, defined medium is highly susceptible to changes in pH resulting from fluctuations in CO_2 pressure. Buffering with Hepes partially alleviates this problem. Even with additional buffering, care must be taken to minimize the time the cultures are outside the incubator, and access to the incubator should be restricted to maintain the proper CO_2 pressure (5%).

Other equipment required includes: micropipettors, haemocytometers and a clinical centrifuge. The researcher should have access to a source of very high purity water, such as thrice glass distilled or from a Millipore Milli-Q ultrafiltration system, since the metal ions and salts present in tap or even deionized water may interfere with the stability of purified growth factors (for discussion see ref. 1). Standard disposable radiation-sterilized plasticware and pipette tips should be used; no glassware or plasticware should be sterilized by autoclaving, as that exposes them to tap water steam. Similarly, solutions must all be sterilized by passage through 0.2 μm filters and stored in disposable plasticware. Solutions and growth factor that are to be stored at −20°C should not be stored in frost-free freezers, as the periodic warming cycle such freezers use to avoid ice build-up will reduce shelf life of the samples.

Nutrient media, phosphate-buffered saline (PBS) and salt solutions may be stored in polycarbonate flasks. All growth factor containing solutions should be stored in polypropylene tubes, as some purified proteins will adhere to polycarbonate. All plasticware used for the handling of cells should also be polypropylene, since cells may attach to polycarbonate.

2.2 **Required solutions**

2.2.1 *Storage and handling*

Except where noted, all solutions should be aliquoted to single use amounts and stored at −20°C, to prevent breakdown of proteins and to preserve the buffering capacities of the solutions. While some solutions can be refrozen once or twice, repeated freezing and thawing should be avoided. Generally the amount needed on a given day is thawed, used and the excess discarded. Thus nutrient media should be frozen in 100 to 200-ml aliquots, while growth factor solutions should be aliquoted into the amounts required for 25–50 dishes. We discard frozen stocks after 6 months of storage.

2.2.2 *Recipes of media and media components*

When purchasing chemicals and growth factors, obtain tissue culture grade when available.

Antibiotics (obtained from Sigma): ampicillin, sodium salt, 2.5 mg/ml; penicillin-G, sodium salt, 12 mg/ml; streptomycin sulphate, 20 mg/ml. Make each antibiotic up in water, filter sterilize, and store in 20 ml aliquots.

Selenium 10^{-5} M: dissolve 1.7 μg/ml sodium selenite (Sigma) in water, filter sterilize, and store in 2-ml aliquots.

F12/Dulbecco's modified Eagle's medium (DMEM) with Hepes 10 mM, selenium 10^{-8} M (F12/DMEM): to one package, F12 Ham's nutrient mixture (Gibco) and 1 package DMEM, high glucose, without sodium pyruvate or sodium bicarbonate (DMEM, Gibco) add 9 g Hepes (Sigma), 20 ml each of ampicillin, penicillin and streptomycin stocks and 4.8 g sodium bicarbonate. Bring to 2 litres with water, add 2 ml selenium 10^{-5} M, filter and store in 200-ml aliquots.

Phosphate-buffered saline: combine 200 mg KCl, 200 mg KH_2PO_4, 8 g NaCl and 2.16 g $Na_2HPO_4 \cdot 7H_2O$. Bring to 1 litre with water. Filter and store in 100-ml aliquots.

Soybean trypsin inhibitor (STI, Sigma): 1 mg/ml in F12/DME, filter sterilize and store in 2-ml aliquots.

Transferrin, human, iron free (Sigma): 2.5 mg/ml in F12/DME, filter sterilize and store in 1 ml aliquots.

Insulin, bovine pancreas (Sigma): 1 mg/ml in water, to which is added 0.001 vol. of 1 M HCl (final concentration 0.001 M HCl), filter sterilize, and store in 1-ml aliquots.

Epidermal growth factor, mouse (EGF, from Collaborative Research): 100 μg/ml in water, filter sterilize and aliquot 10 μl per tube. Just before use, add 1 ml of F12/DMEM per tube (final concentration 10 μg/ml).

Platelet-derived growth factor, human (PDGF): obtained from Collaborative Research in vials of 50 units and stored at $-70°C$. Just before use thaw and suspend in 1 ml of F12/DMEM. Note: human PDGF consists of a heterodimer of A and B chains (8). PDGF is also available from other species, at reduced price, but these do not necessarily have the same oligomeric structure; porcine PDGF, for instance is composed of a homodimer (9). There is some evidence that PDGF homodimers and heterodimers may bind to different receptors, and have different biological effects (8). On the other hand, some researchers find that in their system porcine and human PDGF are interchangeable. Since it is not clear what effect oligomeric structure has on PDGF function the researcher should determine empirically how PDGF from a particular source functions in their system, and then be consistent in their use of that source.

Fibronectin, human plasma: obtained from Chemicon or Collaborative Research. Dilute to 1 mg/ml with PBS and store in 1-ml aliquots. When ready to use, thaw at 37°C. Sometimes fibronectin will form a flocculant precipitate upon thawing. In this case, be sure to thoroughly resuspend the precipitate by pipetting or vortexing before use.

Fibroblast growth factor, basic, bovine (FGF): obtained from Collaborative Research in 10-mg vials. Resuspend in 1-ml PBS (10 mg/ml) just before use.

Bovine serum albumin fraction V (BSA, Sigma): 10 mg/ml in water, filter sterilize and store in 100-μl aliquots.

BSA/human high density lipoprotein (BSA/HDL): HDL is obtained as a 10 mg/ml solution from Bionetics. Store at 4°C. Just before use prepare by adding to 1 ml (of HDL) 20 μl of BSA 10 mg/ml (final concentration HDL 10 mg/ml, BSA 200 μg/ml). HDL is relatively unstable, and should not be stored for more than 6 weeks after purchase.

Poly-D-lysine (Sigma): 100 mg/ml in PBS, filter sterilize and store in 0.5-ml aliquots. Poly-D-lysine can be purchased as endotoxin free, tissue culture tested, at considerable increase in cost. If the relatively inexpensive untested reagent is used, it is imperative that it be tested in the laboratory for toxicity. We have found that some lots of the untested variety show toxicity at the concentration we use of 1 mg per dish, manifested by the cells attaching to the dish but not spreading, remaining as a ball.

In addition the researcher will need the following materials for standard cell culture in serum supplemented media, which need not be made with high purity water and may be treated as usual: DMEM, minimal essential medium (MEM) from Gibco, calf serum and foetal bovine serum (FBS) (Hyclone), PBS with EDTA 0.5 mM (PBS/EDTA) and PBS/EDTA with 0.1% trypsin (Gibco) (PBS/EDTA/trypsin).

2.2.3 *Recipes for solutions for transfection*

2 × HBS: Combine 1 g of Hepes and 1.6 g of NaCl in 80 ml of deionized water. Adjust pH to 7.10 ± 0.05 with 2 M NaOH. Bring the volume to 100 ml with water. Filter sterilize. Store at room temperature for no more than 1 month.

70 mM NaH_2PO_4 (monobasic): dissolve 0.96 g of NaH_2PO_4 (Fisher) in deionized water to a final volume of 100 ml. Filter sterilize.

70 mM Na_2HPO_4 (dibasic): Dissolve 0.99 g Na_2HPO_4 (Fisher) in deionized water, to a final volume of 100 ml. Filter sterilize.

2 M CaCl$_2$: dissolve 29.4 g CaCl$_2$·2H$_2$O (Mallinkrodt) in deionized water to a final volume of 100 ml. Filter sterilize and aliquot in 1–5-ml aliquots. Store at −20°C.

Salmon sperm DNA, 1 mg/ml, sheared: open a fresh vial of salmon sperm DNA, sodium salt. Using sterile technique and sterile forceps, transfer the DNA to a sterile dish and weigh. Transfer to a 50-ml polypropylene tube and add sterile deionized water to make a solution of 1 mg/ml. Because of the fibrous nature of the DNA it is difficult to weigh an exact amount; thus it is easier to obtain an approximate amount and vary the amount of water added to get the proper concentration. Vortex vigorously until the DNA is completely dissolved. The solution will be very viscous. Pour the solution into a 50-ml syringe with an 18-gauge needle and eject forcefully, shearing the DNA. Repeat passing the solution through the needle 3–5 times, until most of the viscosity has been lost. Aliquot into 1-ml aliquots and freeze.

2.2.4 *Staining solutions*

Crystal violet, 0.1%: dissolve 0.1 g of crystal violet in 100 ml of 20% ethanol. Store in a foil covered bottle.

Giemsa, 10%: dissolve 1.0 g of Giemsa stain in 100 ml of 50% ethanol and 50% acetone. Store in a foil covered bottle.

2.3 **Cell lines**

The NIH3T3 (10, 11) cell line used in the experiments described below were single cell cloned from a line obtained from Dr Robert Weinberg (MIT) and are maintained in DME supplemented with 5% calf serum, penicillin and streptomycin. C3H 10T1/2 (12) were single cell cloned from a line obtained from the American Type Culture Collection, Rockville, Maryland, USA (ATCC), and maintained in MEM supplemented with 10% foetal bovine serum, penicillin and streptomycin.

Great care should be taken in the treatment of fibroblast cultures to maintain a precise passage schedule, as small variations in passage parameters can cause the characteristics of the line to change over a short period of time. All 3T3 lines should be passed every 3 days, inoculating 5-cm^2 dishes at a density of 3 × 10^5 per dish (10, 1), consistent with the original 3T3 protocol used to isolate the line. For 75-cm^2 flasks use 1 × 10^6 cells per 75-cm^2 flasks (1.5 × 10^4 cells/cm^2). All 10T1/2 lines should be passed every 10 days at a density of 0.5 × 10^5 cells per 6-cm dish (12), changing the medium at 5 days (1.8 × 10^3 cells/cm^2, 1 × 10^4 cells/75-cm^2 flask). When a new line is obtained it should be passed to a large number of flasks using the above procedure and frozen in aliquots of 3 × 10^5 for 3T3 or 0.5 × 10^5 for 10T1/2 cells. Prior to each set of experiments the cells should be thawed, passaged until the cells needed are obtained and then used. A fresh aliquot should be thawed every 2–3 weeks.

2.4 **Plasmids**

To obtain some of our transformed lines we cotransfected a test DNA with a selectable marker which confers, upon eukaryotic cells, resistance to the neomycin analogue G418. Various constructs containing the selectable marker gene bacterial aminoglycoside phosphotransferase under varying promoters are available. In these experiments we have used pSHL72 (S. H. Larsen, personal communication) which uses the herpes virus thymidine kinase promoter and pSV2neo (13) which uses the SV40 enhancer and promoter.

3. GENERAL PROCEDURES

3.1 **Notes**

Cell lines are initially maintained in the appropriate serum-supplemented medium. Prior to the experiment, the cells needed are cultured in 75-cm² flasks to subconfluence (5×10^5 cells per flask, 6.7×10^3 cells/cm²). On the day the cells are to be plated into defined medium, 3.5-cm dishes are labelled, precoated with attachment factors, filled with 1 ml of unsupplemented nutrient medium and equilibrated in the incubator for at least 30 min. The cell monolayers are then washed and dislodged prior to seeding in 35-cm dishes. Some lines (specifically our line of NIH3T3 cells) can be easily dislodged with the application of only PBS/EDTA. Most lines will require trypsin to dislodge; in this case they must be treated with STI and BSA, pelleted, washed once and resuspended in medium. After being dislodged, the suspensions are counted and diluted in media to 1×10^4 cells/ml; 1 ml (1×10^4) of cells is then added to each dish, bringing the total medium volume to 2 ml. The dishes are put back into the incubator for at least 30 min, to allow the cells to attach and the media to equilibrate to the proper temperature and pH. Appropriate growth factors are then added individually with a micropipettor, and the dishes are returned to the incubator. When using a new cell line whose growth in defined media has not been characterized, one should also include dishes which are supplemented with serum.

Every 3–4 days the medium is removed and new nutrient media (2 ml/dish) is applied. Growth factors are again added individually.

Many assays conclude with counting the cells on each dish. If so, then the medium is removed, the dishes washed and the cells dislodged with 1 ml PBS/EDTA/trypsin. The cell suspensions are then counted with a haemocytometer. It is suggested that in any experiment all treatments be performed in duplicate and all counts be done in duplicate. Thus each data point represents an average of four haemocytometer counts; two counts for each of two dishes.

Most experiments involve many different combinations of supplemented growth factors, perhaps at different concentrations, and thus the researcher must add growth factors individually to each dish as described. However, some experiments may call for one or more supplements to be added to all dishes. In this case, supplements may be added to a quantity of medium in polypropylene tubes immediately before plating.

In all procedures, special care should be taken to minimize the time dishes spend outside the incubator.

3.2 General protocols for culturing in defined media

3.2.1 *Precoating with fibronectin*

  (i) Label the required number of dishes. Thaw 1 ml of fibronectin (1 mg/ml) in a 37°C water bath, and dilute 1:50 into PBS in a polypropylene tube.

 (ii) Add 1 ml of the diluted fibronectin (0.02 mg/ml) to each 3.5-cm dish, swirling each dish after addition to spread the solution over the entire surface. Place in the incubator for 2 h.

(iii) Aspirate off the fibronectin. Add 1 ml of PBS to each dish, swirl and aspirate off. Add 1 ml of unsupplemented F12/DME to each dish and return dishes to the incubator.

3.2.2 *Precoating with poly-D-lysine and fibronectin*

  (i) Label the required number of dishes. Thaw poly-D-lysine in 37°C waterbath. Add 1 ml of PBS to each dish. Swirl each dish to cover the entire surface with liquid. Add 10 μl of poly-D-lysine (100 mg/ml) to each dish (final concentration 1 mg/ml), swirling after each addition. Place dishes in the incubator for 2 h.

 (ii) Thaw 1 ml of fibronectin (1 mg/ml), and dilute 1:50 into PBS in a polypropylene tube. Remove the dishes from the incubator and aspirate off the poly-D-lysine.

(iii) Add 1 ml of the diluted fibronectin (0.02 mg/ml) to each 3.5-cm dish, swirling each dish after addition to spread the solution over the entire surface. Place in the incubator for 2 h.

(iv) Aspirate off the fibronectin. Add 1 ml of PBS to each dish, swirl and aspirate off. Add 1 ml of unsupplemented F12/DMEM to each dish and return the dishes to the incubator.

3.2.3 *Dislodging cells with PBS/EDTA*

  (i) Obtain a 75-cm^2 flask of cells in serum supplemented medium. Aspirate the medium from flask. With the flask upright on its end, add 5 ml of PBS/EDTA. Lay the flask flat and swirl gently but quickly, washing the monolayer. (Note: since NIH3T3 cells detach easily in PBS/EDTA, avoid leaving the wash on an NIH3T3 monolayer for more than several seconds.) Aspirate off the liquid.

 (ii) Add 3 ml of PBS/EDTA. Place in incubator for 10 min. Knock firmly against the side of a table to dislodge cells.

3.2.4 *Dislodging cells with PBS/EDTA/trypsin*

  (i) Obtain a 75-cm^2 flask of cells in serum supplemented medium. Aspirate the medium from the flask. With the flask upright on its end, add 5 ml of PBS/EDTA. Lay the flask flat and swirl gently but quickly, washing the monolayer. Aspirate off the PBS/EDTA.

(ii) Add 1.5 ml of PBS/EDTA/trypsin. Swirl to cover the monolayer. Set flat either at room temperature or in the incubator for 5 min. Knock firmly to dislodge cells. (Note: the amount of trypsin used and the time required may vary from cell line to cell line, and should be determined empirically.)

(iii) Aliquot to a 15-ml polypropylene tube 3 ml of STI, $100\,\mu l$ of BSA and 2 ml of unsupplemented F12/DMEM medium. Add 2 ml of medium to the flask, transfer the cell suspension to the tube and mix by gentle pipetting. Pellet in a clinical centrifuge for 5 min at 2000 r.p.m.

(iv) Aspirate off the supernatant. Resuspend cells by gently pipetting in 5 ml of F12/DME. Again, pellet for 5 min.

(v) Aspirate off the supernatant. Resuspend cells in 8 ml of unsupplemented F12/DMEM.

3.2.5 *Changing the media*

(i) Aspirate off the media. Add 2 ml of F12/DMEM to each dish. If this process takes longer than 10 min (such as when many dishes are used), return the dishes to the incubator for 15–30 min to equilibrate temperature and pH. Otherwise proceed to the next step.

(ii) Add individual growth factors to the dishes using a micropipettor, as the experiment dictates. Swirl the dishes after each addition. Return the dishes to the incubator.

4. DEFINED MEDIA PROTOCOLS FOR NIH3T3 AND C3H 10T1/2 CELLS

4.1 General notes on NIH3T3 in defined media

NIH3T3 cells are normally maintained in DME supplemented with 5% calf serum. It is particularly important when maintaining this line not to allow it to grow to confluence, as spontaneous variants frequently arise when the cells are confluent. They will proliferate in F12/DMEM supplemented with transferrin ($25\,\mu g/ml$), insulin ($10\,\mu g/ml$), EGF (100 ng/ml), fibroblast growth factor (FGF) (100 ng/ml) and PDGF (0.5 U/ml) on dishes precoated with poly-D-lysine and fibronectin.

Our cell line detaches quite easily from plastic in the presence of EDTA. Thus care must be taken to avoid washing the cells away when initially washing the monolayer. However, this makes it possible to dislodge the cells without the use of trypsin and consequently to avoid the extra stress of removing the trypsin with STI, pelletting and washing. While you should still use trypsin in routine passing in serum-supplemented medium, avoid its use if possible when passing cells to serum-free medium.

4.1.1 *Passing from serum-containing medium to defined medium*

(i) Obtain a 75-cm^2 flask of NIH3T3 cells in DMEM, 5% calf serum. Precoat 3.5-cm dishes with poly-D-lysine and fibronectin, as in Section 3.2.2. Place the dishes in the incubator.

(ii) Dislodge the cells from the flask. If the cell line will dislodge using only PBS/EDTA, use the procedure given in Section 3.2.3. Otherwise use the procedure given in Section 3.2.4. Bring the volume of the cell suspension to 8 ml with F12/DMEM.

(iii) Count with a haemocytometer. Dilute into F12/DMEM to a cell density of 1×10^4 cells/ml. Remove the precoated dishes from the incubator. Add 1 ml of cell suspension to each dish. Return the dishes to incubator for 30–60 min.

(iv) Thaw and prepare the required growth factors (see below). Remove the dishes from the incubator. Add to each dish 20 μl of transferrin, insulin, EGF, FGF and PDGF stock solutions. Return dishes to incubator.

4.2 General notes on C3H 10T1/2 in defined media

C3H 10T1/2 cells are routinely carried in MEM supplemented with 10% foetal bovine serum. They will proliferate in defined media in the presence of transferrin, insulin, EGF, and BSA/HDL on fibronectin-coated dishes. Interestingly, in our system they do not proliferate at a higher rate in the presence of PDGF.

4.2.1 *Passing from serum-supplemented medium to defined medium*

(i) Obtain a 75-cm² flask of C3H 10T1/2 cells in MEM, 10% foetal bovine serum. Precoat 3.5-cm dishes with fibronectin, as in Section 3.2.1. Place the dishes in the incubator.

(ii) Dislodge cells from the flask with PBS/EDTA/trypsin using the procedure given in Section 3.2.4. Bring the volume of cell suspension to 8 ml with F12/DMEM.

(iii) Count with a haemocytometer. Dilute into F12/DME to a cell density of 1×10^4 cells/ml. Remove the precoated dishes from the incubator. Add 1 ml of cell suspension to each dish. Return the dishes to the incubator for 30–60 min.

(iv) Thaw and prepare the required growth factors. Remove the dishes from the incubator. Add to each dish 20 μl of transferrin, insulin and EGF and 2 μl of BSA/HDL stock solutions. Return the dishes to the incubator.

5. MISCELLANEOUS PROCEDURES

Many of the experiments in Section 6 assume that a cell line has been obtained that is different from the parental cell line, often because it is expressing a transforming gene. Such a cell line could be generated by transfecting normal fibroblasts with a known transforming gene, picking the resulting foci, and generating a cell line from a focus. Alternatively a plasmid containing the gene of interest can be cotransfected with a selectable marker, such as pSV2neo, which confers resistance to G418 on cells. In this case G418 resistant lines are then generated which express the cotransfected gene in the absence of any selection for that gene. Here we provide a few basic methods concerned with the generation of cell lines to be tested in defined media.

5.1 **General notes on transfection**

The following is a modification of the calcium phosphate transfection procedure of Graham and van der Ebb (14). It involves preparing a solution of 20 mM Hepes, 150 mM NaCl, 0.7 mM sodium phosphate, 125 mM $CaCl_2$, 20 μg/ml DNA, pH 7.1. The DNA includes the plasmid of interest added to carrier DNA to make the total DNA concentration 20 μg/ml. Preferably homologous DNA isolated from the recipient cell line should be used as carrier DNA (for instance, NIH3T3 DNA for transfection into NIH3T3 cells). Alternatively, sheared salmon sperm DNA which has been traditionally used as carrier works well. This is added to a culture of actively dividing fibroblasts and allowed to adsorb to the cells. After 4 h excess precipitate is removed and the cells exposed to an osmotic shock with 20% glycerol (15).

Note: it is convenient to prepare DNA for use in transfection on the day before (or earlier) by aliquoting the appropriate amount to an Eppendorf tube, ethanol precipitating the DNA and resuspending it in sterile water. This has the multiple advantages of simplifying the procedure on the day of the transfection, ensuring that the DNA is sterile and ensuring that the DNA solution does not contain other solutes.

5.1.1 *Transfection of NIH3T3*

(i) On the day before the transfection, plate 10-cm dishes with the recipient line, 5×10^5 cells per dish.

(ii) On the day of transfection, prepare 0.5 ml of solution A for each transfection. For example, for ten transfections prepare 5 ml by adding 50 μl each of monobasic and dibasic 70 mM phosphate buffers to 5 ml of $2 \times$ HBS.

(iii) For each transfection prepare a solution B, by combining 88 μl of 2 M $CaCl_2$, the transfection DNA (1–5 μg), carrier DNA (1 mg/ml) to make the total DNA concentration 10 μg, and deionized sterile water to a total volume of 0.5 ml.

(iv) Add 0.5 ml of solution A to each solution B. Add using a Pasteur pipette, swirling the tube while adding A to B.

(v) Set at room temperature for 20 min.

(vi) Remove medium from the 10-cm dishes. Add transfection cocktails to the dishes (1 ml per dish). Set the dishes in the incubator for 1 h, rocking the dishes to distribute the precipitate every 15 min.

(vii) Add 5 ml of serum-supplemented medium to each dish and return the dishes to the incubator for 4 h.

(viii) Prepare 2 ml of 20% glycerol in medium per dish. Remove the dishes from the incubator and aspirate the medium from three dishes at a time. Add 2 ml of 20% glycerol to each dish, swirl to cover. Let sit for exactly 2 min, then add 8 ml of unsupplemented medium to each dish. Swirl and aspirate off. Wash dishes with another 5 ml of unsupplemented medium, and finally add 10 ml of serum-supplemented medium. Return dishes to incubator.

(ix) After 2 days recovery time the cells may be assayed or passed into selective media.

5.1.2 *Transfection of C3H 10T1/2*

(i) 24 h before the transfection, plate 10-cm dishes with C3H 10T1/2 at 5×10^5 cells per dish in MEM supplemented with 10% FBS. 4 h before transfecting, remove the medium and replace with 10 ml of MEM with 20% FBS.

(ii) On the day of transfection, prepare 0.5 ml of solution A for each transfection. For example, for ten transfections prepare 5 ml by adding 50 μl each of monobasic and dibasic 70 mM phosphate buffers to 5 ml 2 $\times$ HBS.

(iii) For each transfection prepare a solution B, by combining 88 μl of 2 M CaCl$_2$, the transfection DNA (1–5 μg), carrier DNA (1 mg/ml) to make the total DNA concentration 20 μg, and deionized sterile water to a total volume of 0.5 ml.

(iv) Add 0.5 ml of solution A to each solution B. Using a Pasteur pipette, add A to B a drop at a time.

(v) Set at room temperature for 20 min.

(vi) Add the transfection cocktails to the medium in the dishes (1 ml per 10-cm dish). Set the dishes in the incubator for 4–5 h. Alternatively, incubate for 15 h with no glycerol shock.

(vii) Prepare 5 ml of 20% glycerol in 1 $\times$ HBS per dish. Remove the dishes from the incubator and aspirate the medium from three dishes at a time. Wash the dishes twice with 5 ml of MEM supplemented with 10% FBS. Add 5 ml of 20% glycerol to each dish, swirl to cover. Let sit for exactly 1 min, then aspirate off the glycerol solution. Let sit 1 more min, and wash with 5 ml 1 $\times$ HBS. Add 10 ml of MEM supplemented with 10% FBS. Return the dishes to the incubator.

(viii) After 2 days recovery time cells may be assayed or passed 1 : 3 into selective medium.

5.2 General notes on focus assay

Normal fibroblasts generally exhibit density dependent inhibition of proliferation, meaning that they will grow on a plastic surface until the entire surface is covered with a single layer of cells (a monolayer) and then stop proliferating. Most transforming genes will release expressing cells from this inhibition, and thus will cause piles of cells (foci) to arise on a confluent monolayer.

The focus assay can be used to assay a transfected line for the presence of transformed cells, or as a way to generate cell lines expressing transforming genes for further study.

5.2.1 *Focus assay*

(i) Trypsinize a flask of cells and plate 5×10^5 cells on each of two 10-cm dishes in media supplemented with calf serum.

(ii) Maintain the cultures in the incubator for 4 weeks, changing the media every 3 days. To change the medium aspirate off the medium, being careful not to disturb the cell layer with the tip of the pipette. Add 10 ml of fresh medium with a wide bore pipette, holding the tip of the pipette against the inner surface of the side of the dish and slowly letting the medium run down the side of the dish. Even with these precautions the force of pipetting media onto the monolayer can sometimes lift a section of monolayer off the dish, which then folds over and reattaches, forming a pseudofocus. Thus it is useful to place a mark on the side of the dish, and to try to add the media on the same place on the dish each time. Then any unusual aberations in the monolayer morphology near the mark can be disregarded.

(iii) Examine the dishes with the naked eye and under low power magnification for piles of cells at 2, 3 and 4 weeks, respectively.

(iv) After foci have arisen the foci can be picked with a cloning ring or scraped off with the tip of a Pasteur pipette and transferred to a microwell containing PBS/EDTA/trypsin, allowed to dissociate, and transferred into a small flask containing medium. Alternatively the dishes can be stained with 1% Giemsa stain in ethanol and photographed.

5.3 Picking colonies

We use two methods of picking colonies in our laboratory: cloning rings and picking the colony with a Pasteur pipette. The former involves placing a ring greased with sterile silicon grease around the colony and adding trypsin solution to the inside of the ring. This method is relatively quick, but requires some experience to completely seal the edge of the ring around the colony, and requires the preparation of sterile rings and grease. Picking with a pipette involves physically scraping the cells of the colony off the dish surface and transferring to a separate dish for trypsinization and dissociation. This latter procedure requires more manipulation, but is in some ways more foolproof.

After a colony or focus is picked a single cell cloned line can be derived and used for various assays. For some assays single cell cloning, while usually desirable, may not be necessary, since the vast majority of the cell population is derived from the colony or focus. In this case a line may be grown directly from the colony or focus and assayed.

5.3.1 *Picking colonies with cloning rings*

(i) Cloning rings are small hollow cylinders open at both ends. Glass cloning rings are commercially available. In our laboratory we find that cloning rings may be easily made from truncated Eppendorf tubes, and these rings are easier to use than purchased rings. Prepare cloning rings from 1.5 ml Eppendorf tubes by heating a scalpel over a gas burner and cutting off the bottom two-thirds of the tube and the cap. Autoclave the rings and a small quantity of silicon vacuum grease, and then fill a 10-ml syringe with the sterile grease.

(ii) Remove the medium from the dish and wash the cells with 2 ml of PBS/EDTA. Holding the ring with sterile forceps apply a thin line of grease over the smooth edge of the cloning ring (formerly the top edge of the Eppendorf tube). Set the ring down over the colony, twisting slightly to seal the edge against the bottom of the dish.

(iii) Add about 0.2 ml of PBS/EDTA/trypsin to the inside of the ring. Allow to trypsinize for 5 min. Remove the cell suspension with a Pasteur pipette, pipetting in and out very gently if necessary to dislodge the cells. Transfer the cell suspension to a small flask or dish containing serum-supplemented medium.

5.3.2 *Picking colonies with a Pasteur pipette*

 (i) Aspirate media from the dish. Add 1 ml of PBS/EDTA/trypsin to a microwell or 3.5-cm dish.
 (ii) Draw up a few drops of PBS/EDTA/trypsin into a Pasteur pipette. Using either the tip of the pipette, a cell scraper, or a sterile flat toothpick, scrape around the focus or colony until it is dislodged as a ball of cells. Use the liquid in the pipette to draw the ball of cells into the pipette. Expel into the PBS/EDTA/trypsin. Allow to trypsinize for 5 min.
(iii) Dissociate by rapid pipetting several times, then transfer 0.5 ml of the cell suspension to a flask containing 15 ml of serum-supplemented medium. Culture as usual to obtain a cell line.

5.3.3 *Single cell cloning*

 (i) Trypsinize the culture of cells as if you were going to pass them. Count the cell density with a haemocytometer.
 (ii) Prepare two 10-ml dilutions in polypropylene tubes using serum-supplemented media, containing 100 and 10 cells, respectively.
(iii) For each dilution, aliquot 0.1 ml into each microwell of a 96-microwell plate. Thus three microwell plates will be prepared.
(iv) Culture for 4 days. Examine each plate at 40× magnification and count the number of wells which contain cells. Retain the plate in which only 10–20% of the microwells contain cells—discard the other.
 (v) Add another 0.1 ml of supplemented medium to the microwells and culture another 4 days. Trypsinize and pass to a small dish.

5.4 **Staining dishes with crystal violet or Giemsa stain**

Often it is desirable to have photographs of an entire dish to visually show colonies, foci or unusual variations in a cell monolayer. We use the darkly staining crystal violet to stain colonies of cells against the background of a bare dish or to stain dishes with different cell densities. Crystal violet does not work well when staining foci, since it stains the background cell monolayer too darkly. For staining foci or other variations in a cell monolayer Giemsa stain should be used.

To stain with either stain, pour off the medium and add 1–2 ml of stain. Rotate to cover the dish surface. Allow to sit for about 30 sec and pour off the stain. Wash in a sink under a gentle stream of water. Set tilted on paper towels to air-dry.

6. EXPERIMENTS

6.1 **Assays of response to growth factors**

6.1.1 *Assay of an oncogene-transformed NIH3T3- or C3H 10T1/2-derived cell line for altered response to growth factors, 7-day assay*

(i) Precoat 24 3.5-cm dishes with poly-D-lysine and fibronectin as in Section 3.2.2 for NIH3T3, or with fibronectin as in Section 3.2.1 for C3H 10T1/2.

(ii) Pass NIH3T3 or C3H 10T1/2 cells to 12 dishes as in Section 4.1.1(ii)–(iv) or Section 4.2.1(ii)–(iv). However, instead of adding all factors to all dishes, add to pairs of dishes all combinations of growth factors with each group missing one of the required factors. For instance, for NIH3T3 cells the following combinations might be used.

> Group 1: transferrin.
> Group 2: transferrin, insulin, EGF, FGF.
> Group 3: transferrin, insulin, FGF, PDGF.
> Group 4: transferrin, EGF, FGF, PDGF.
> Group 5: transferrin, insulin, EGF, PDGF.
> Group 6: transferrin, insulin, EGF, FGF, PDGF.

Such a set of combinations tests the growth of the cell line in media lacking insulin, EGF, FGF or PDGF.

(iii) Pass transformed cells to 12 dishes and supplement as in (ii).

(iv) After 4 days remove the media, add 2 ml of fresh F12/DMEM and supplement as in (ii).

(v) After 3 more days (total of 7 days), remove media and add 1 ml of PBS/EDTA/trypsin. Set the dishes at room temperature for 5 min. [Cells on dishes precoated with poly-D-lysine are somewhat resistant to normal trypsinization procedures. Therefore double the trypsin concentration and incubate the trypsinizing cells in the incubator for 10 min when using NIH3T3-derived lines. Alternatively, it has been reported that the use of poly-L-lysine instead of poly-D-lysine facilitates trypsinization (16).] Dislodge cells by rapid pipetting and count with a haemocytometer. Compare growth of the transformed line to the parental line in each combination of supplements.

6.1.2 *Titrate requirement for a growth factor*

(i) Precoat 10 3.5-cm dishes with a poly-D-lysine and fibronectin as in Section 3.2.2 for a NIH3T3 derived line, or with fibronectin as in Section 3.2.1 for a C3H 10T1/2 derived line.

(ii) Pass NIH3T3 or C3H 10T1/2 cells to ten dishes as in Section 4.1.1(ii)–(iv) or as in Section 4.1.2(ii)–(iv). However, instead of adding all factors to all dishes, vary the concentration of the growth factor being tested over a 1000-fold range. For instance, to assay an NIH3T3 line for response to FGF one might vary the FGF concentration in the following manner: 0.1 ng/ml, 1 ng/ml, 10 ng/ml, 100 ng/ml, 1 μg/ml.

(iii) After 4 days remove the media, add 2 ml of fresh F12/DMEM, and supplement as in (ii).

(iv) After 3 more days (total of 7 days), remove media, add 1 ml of PBS/EDTA/ trypsin and set at room temperature for 5 min [see Section 5.1.1(v)]. Dislodge cells by rapid pipetting, and count with a haemocytometer.

(v) Choose the lowest concentration that induces the maximal cell number. Repeat the experiment, this time bracketing that supplement concentration over a 50- or 100-fold range. Using the results thus obtained, graph the log of the cell density versus growth factor concentration or calculate cell doublings and plot these doublings versus growth factor concentration.

6.2 Selection of transformed cell lines in defined media

6.2.1 *Rationale*

Culture of cells in suboptimal media formulations (i.e. formulations which do not support a maximum rate of growth, such as when required factors are deleted from the medium) places a selective pressure on the cells which would favour some variants which have alterations in their growth control pathways. Thus one should be able to use growth in such media formulations to identify genes involved in those pathways. Such experiments could be performed using media supplemented with only trace amounts of serum. However, in this case, the reduced serum supplement reduces not only the concentrations of all growth factors, but also reduces the concentrations of the trace elements, vitamins and other nutrients also provided by the serum. Defined media allows one to perform similar experiments but with a greater control on the specific selective pressures involved, by allowing the researcher to control the concentration of each individual nutrient and growth factor.

We have used defined media to attempt to identify oncogenes in DNA isolated from human colon carcinoma (6). In our protocol, NIH3T3 cells are transfected with DNA from tumours and the resulting pool of cells is maintained in defined media missing one or two of the required growth factors. This limits the proliferation of the cell population. Against this background of slowly growing cells, colonies of actively proliferating cells sometimes arise. When these colonies are picked and assayed for growth in defined media, they are found to have either reduced requirements for one of the added growth factors or have become much more proliferative in the defined media generally. Presumably, these colonies have arisen from cells that, in the original transfection, took up a fragment of exogenous DNA with a gene which somehow directly or indirectly perturbs the growth control pathways. One can confirm this supposition by repeating the experiment using DNA isolated from the proliferating colony to determine if the phenotype can be transferred through several passages and whether the presence of specific DNA fragments is correlated with the passage of the phenotype.

Such a protocol could be used with DNA from various sources, including tumours or variant cell lines with different growth characteristics. Alternatively, it could be used with cells treated with a mutagen and then grown in defined media to select those that have gained mutations in genes in the growth control pathways. Our procedure of transfecting with tumour DNA is described in Section 7.4. Whatever the source of the cells, the following selection protocol can be used to isolate proliferating cell lines from the initial pool.

6.2.2 *Procedure*

(i) Precoat 10–20 dishes with fibronectin as in Section 3.2.1. Do NOT precoat with poly-D-lysine. We find that discrete, proliferating colonies are more easily identified if the dishes are not precoated with poly-D-lysine even though such conditions are suboptimal for growth. It has been reported that fibroblasts exhibit enhanced motility on dishes coated with poly-D-lysine (17). We suspect that the increased movement of cells away from a colony's centre is responsible for the poor definition of colonies on poly-D-lysine-coated dishes.

(ii) Pass the transformed line into defined media as in Section 4.1.1, adding supplements as follows: to two dishes, add all factors (transferrin, insulin, EGF, FGF and PDGF). To the remaining dishes add all but one or two growth factors, leaving either insulin, EGF, FGF and/or PDGF out of the media.

(iii) Change the media every 3–4 days for a total of 14 days.

(iv) After 14 days, examine the selective dishes under the microscope for the presence of discrete, circular colonies of actively proliferating cells against a background of slowly growing cells. A colony is functionally defined as a group of some minimum number of cells, each within one cell diameter of another cell in the colony, against a background of sparser and more randomly distributed cells. The minimum number of cells in a colony will depend on the number of days (and anticipated doublings) since plating. Typically a colony will be first recognizable after about 1 week, when it should have from 16–32 (or more) cells. After 14 days a colony should have from 10^2 to 10^3 cells. Pick the colony and transfer to a flask containing serum-supplemented medium. Alternatively, use a cell scraper to remove all cells except those in the colony from the dish. Amplify the culture and assay for changes in growth factor requirements as in Section 4.1.2.

6.3 Modification of a media formulation for a variant cell line

When starting defined media experiments with a cell line, the media formulation may have to be optimized for that cell line. This is even necessary if the researcher is using the NIH3T3 or C3H 10T1/2 lines, since various isolates of these lines obtained from different sources have different characteristics. Once the researcher has tested the line in the given medium formulation and decides that the formula must be modified, the following general approach may be used. First, if desired, other growth factors or nutrients not used in the standard medium may be tested for effect, using a 7-day assay similar to Section 6.1.1. Ideas on which factors and nutrients could be tried can be obtained from the several reviews on the subject (1–3, 18). Secondly, each growth factor now included in the media should be titrated as in Section 6.1.2 to determine the optimal concentration. Generally an optimal concentration is chosen such that it falls within the range where a graph of cell doublings versus factor concentration changes from linear to asymptotic and induces a number of cell doublings within 5 or 10% of that induced by the maximum factor concentration. Thirdly, using a medium with the new factor concentrations thus determined, each optimal concentration must again be titrated, to determine if the response to any supplement has been altered by changes in concentration of other supplements. This last step is required because frequently responses to a group of growth factors can not be predicted from responses to individual factors and vice versa. Rather, factor-induced pathways interact with each other in an unpredictable manner. Thus when any change in the media formulation is made, all other elements of the formulation must be rechecked.

7. EXAMPLE RESULTS

7.1 **7-day assay of growth of NIH3T3 in the presence of pairs of growth factors**

NIH3T3 cells were dislodged by PBS/EDTA (Section 3.2.3) and plated on 16 3.5-cm dishes as precoated as in Sections 3.2.2 and 4.1.1. Pairs of dishes were supplemented as follows, with all serum-free dishes receiving transferrin (I, insulin; E, EGF; P, PDGF; F, FGF): no supplements except transferrin, I + E, I + P, I + F, E + P, E + F, I + E + F + P, 5% calf serum. The media was changed on the 4th day. Dishes were trypsinized and counted on the 7th day.

Counts are the average of two dishes, two counts per dish. We have presented for illustration both the raw data and the final results. Raw counts are presented in *Table 1*. Final results are expressed as cells per dish and as cell doublings per dish (*Table 2*). The number of cell doublings is calculated as:

$$\log[(\text{cell density})/(\text{initial density})] - \log[2]$$

Table 1. Raw data for Section 6.1. Lists haemocytometer counts and averages for each dish. Cell density is calculated by multiplying the haemocytometer count by 1×10^4.

	Dish 1			Dish 2		
Supplements added + transferrin	*Count 1*	*Count 2*	*Average*	*Count 1*	*Count 2*	*Average*
No supplements	1	1	0.06	0	0	0.00
Insulin, EGF	10	20	0.83	18	22	1.11
Insulin, PDGF	22	21	1.19	–	–	–
Insulin, FGF	93	97	5.28	84	122	5.72
EGF, PDGF	3	8	0.31	1	2	0.08
EGF, FGF	8	12	0.55	7	8	0.42
All four	130	105	6.53	36	36	2.00
5% calf serum	1009	1071	120.00	1269	1026	128.00

Table 2. Final results, 7-day assay of NIH3T3, supplemented with pairs of growth factors (Section 6.1).

Supplements added + transferrin	Cell density	Doublings
No supplements	$<1.0 \times 10^4$	<1.0
Insulin, EGF	$<1.0 \times 10^4$	<1.0
Insulin, PDGF	1.2×10^4	<1.0
Insulin, FGF	5.5×10^4	2.5
EGF, PDGF	$<1.0 \times 10^4$	<1.0
EGF, FGF	$<1.0 \times 10^4$	<1.0
All four	4.3×10^4	2.1
5% calf serum	1.2×10^4	7.0

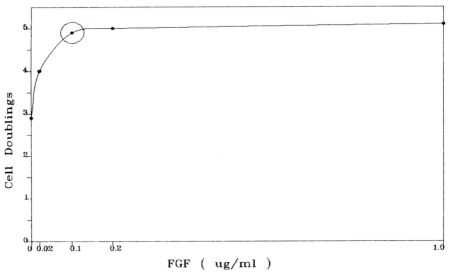

Figure 1. Cell doublings after 7 days as a function of FGF concentration. The experiment was performed as described in Section 6.2. The standard FGF concentration of 100 ng/ml is circled.

or

$$\log[(\text{cell density})/(1 \times 10^4)] - \log[2]$$

(*Note*: the data presented are actually a subset of the experiment described in Section 6.5.)

7.2 Titration of response to FGF by NIH3T3

NIH3T3 were plated on 3.5-cm dishes in media containing standard concentrations of transferrin, insulin, EGF and PDGF, and the following combinations of FGF: 10 ng/ml, 20 ng/ml, 100 ng/ml, 200 ng/ml, 1 μg/ml) using protocols (Sections 3.2.2, 3.2.3, 4.1.1 and 5.1.2). Pairs of dishes were counted after 7 days. Counts represent the average of two dishes, two counts per dish. Results are graphed as cell doublings as a function of concentration (*Figure 1*).

Table 3. Comparison of 10T1/2 and SV40 transformed 10T1/2 indefined media missing either insulin or EFG, 9-day assay (Section 6.3).

	10T1/2		SV40—10T1/2	
Supplements added + *HDL/BSA and transferrin*	*Cell density*	*Doublings*	*Cell density*	*Doublings*
Insulin	1.5×10^4	<1	1.7×10^5	6.9
EGF	2.6×10^4	1.4	1.7×10^6	4.0
Insulin + EGF	4.0×10^5	5.3	1.6×10^6	7.3
10% FBS	6.0×10^5	5.9	7.8×10^5	6.3

Analysis: this is an example of the assays done when developing an optimal defined media for a cell line. Each growth factor must be titrated to give an optimal response. Optimal concentration is chosen in the range where the response curve changes from linear to asymptotic and within one doubling of the maximum cell density. The standard FGF concentration of 100 ng/ml falls within this range.

7.3 Changes in growth factor requirements of C3H 10T1/2 cells after transformation by SV40.

C3H 10T1/2 cells were transfected with pSVB3 (19), a plasmid containing the entire SV40 genome cloned into pBR322 using procedure in Section 5.1.1. Like SV40, pSVB3 induces focus formation on 10T1/2 cells. Foci were picked (Sections 5.2.1 and 5.3.1) and single cell cloned (Section 5.3.3). The resulting cell line was plated onto eight dishes in defined media (Sections 4.2.1 and 5.1.1), along with the parental line 10T1/2. Pairs of dishes were supplemented as follows: insulin, EGF, insulin and EGF, 10% foetal bovine serum (all serum-free dishes received transferrin and BSA/HDL). The media was changed every 3–4 days (this assay was continued to 9 days). After 9 days the dishes were trypsinized and counted. Counts are expressed as cells per dish and as cell doublings (*Table 3*).

Analysis: this experiment demonstrates that this transformed line can proliferate in medium missing either insulin or EGF, yet still responds to the two factors.

7.4 Selection of cells transformed by tumour DNA using defined media

High molecular weight DNA was isolated from samples of human tissue frozen in liquid nitrogen. DNA was transfected into NIH3T3 by the procedure in Section 5.1.1. The transfection used 25 μg high molecular weight DNA and 5 μg of the G418 resistance plasmid pSHL72 per 10-cm dish of cells (note the DNA concentration is 30 μg/ml rather that the 20 μg/ml described in Section 5.1.1). Typically, 20 dishes of cells were transfected with a given DNA sample. Two days after transfection the cells were split 1:3 into fresh dishes and cultured in DMEM supplemented with 5% calf serum containing G418 sulphate (Gibco) 370 active U/ml for 14 days. Using this G418 preselection, all cells which did not take up

DNA in the transfection step are eliminated. We have found that this preselection greatly improves chances for success in the subsequent defined media selection. Also, by counting the number of G418^R colonies, one can determine the efficiency of transfection in terms of the number of independent transfection events. The goal is to have the entire genome of the transfected DNA represented in the pool of cells (i.e. a genomic library). NIH3T3 cells have been reported to take up as

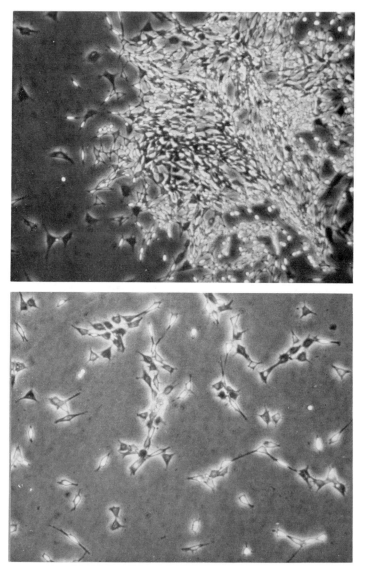

Figure 2. Photomicrograph of proliferating colony. NIH3T3 cells transfected with DNA from lymphocyte no. 1 was maintained for 14 days in defined media supplemented with transferrin, insulin and EGF, as described in Section 6.4. Magnification is 40×. **Top**, a proliferating colony against a background of non-proliferating cells; **bottom**, normal NIH3T3 under the same conditions.

much as 2×10^3 kb of exogenous DNA by calcium phosphate transfection (20). Since the human genome is approximately 2×10^6 kb in size, it is apparent that a genomic library of human DNA in NIH3T3 cells must contain at least several thousand G418[R] colonies to have a reasonable possibility of representing most or all of the human genome.

The resulting pools of cells were then plated in media lacking both FGF and PDGF (containing transferrin, insulin and EGF), and selected for 14 days as in Section 6.2.2. The dishes were scored visually at 40× magnification for proliferating colonies. Colonies were picked and passed into serum-supplemented media. Sample dishes were also trypsinized and counted or stained with 1% violet in 20% ethanol. *Figure 2* shows photomicrographs of portions of example colonies. *Figure 3* shows crystal violet-stained dishes.

In all, we tested six DNA samples isolated from colon carcinomas: two isolated from normal mucosa and three from lymphocytes of patients with Gardner syndrome, a hereditary predisposition to colon cancer. DNA from two carcinoma and all the lymphocyte samples gave rise to colonies in defined media. (Finding colonies in at least two dishes was considered significant.) Cell lines that derived from these colonies were tested for growth in defined media supplemented with transferrin, insulin and EFG in a 14-day assay. The cell line derived from lymphocyte sample number 1 showed significantly increased growth in the 14-day assay (see *Table 4*).

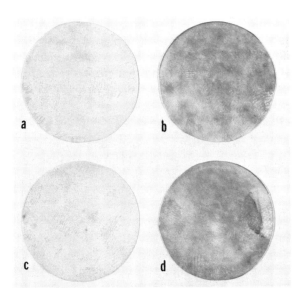

Figure 3. Crystal violet-stained dishes, after selection as in Section 6.4. Normal NIH3T3 cells, **a** and **b**; NIH3T3 transfected with DNA from lymphocyte sample 1, **c** and **d**. Media supplemented with transferrin, insulin and EGF, **a** and **c**; media supplemented with transferrin, insulin, EGF, FGF and PDGF, **b** and **d**.

Table 4. Results of defined media selection on NIH3T3 transfected with DNA isolated from normal colonic mucosa, colon carcinoma and lymphocytes from patients with hereditary predisposition to colon carcinoma (Section 6.4).

Source of DNA	Dishes with colonies/ total dishes	Cell doublings in 14-day assay of derived line in media containing transferrin, insulin, EGF
Normal mucosa No. 1	1/10	–
Normal mucosa No. 2	1/10	–
Carcinoma No. 1	2/10	6.2
Carcinoma No. 2	1/9	–
Carcinoma No. 3	0/10	–
Carcinoma No. 4	0/10	–
Carcinoma No. 5	0/10	–
Carcinoma No. 6	4/10	5.7
Lymphocyte No. 1	8/10	9.0
Lymphocyte No. 2	4/14	3.5
Lymphocyte No. 3	9/10	5.0
	Normal NIH3T3 in 14-day assay	4.2

DNA was isolated from the proliferating colony and the transfection and selection protocol repeated. Colonies were picked and assayed, and DNA was isolated from the cell lines. The DNA was digested with restriction enzymes, separated by agarose gel electrophoresis, transferred to nitrocellulose and the immobilized DNA probed for human sequences using the human *alu* repetitive sequence clone BLUR8. The presence of unique human *alu* sequence-containing *Eco*RI fragments demonstrated that the same phenotype was transferred through a second round via the DNA of the line derived from cells transfected with lymphocyte DNA sample number 1 (6). Comparison of the growth properties of one of these lines, P86, to NIH3T3 cells is described in Section 7.5.

7.5 Comparison of growth in defined media of P86 and NIH3T3

NIH3T3 and P86 cells were grown in serum-supplemented media. 52 3.5-cm dishes were precoated with poly-D-lysine and fibronectin as in Section 3.2.2. Each line was plated to 26 dishes each, as in Section 4.1.1, except supplements were as shown in *Table 5*. The media was changed after 4 days and the dishes counted after 7 days.

Analysis: the parental NIH3T3 line showed significant growth in only media containing transferrin and insulin and FGF, transferrin and all four other growth factors, and with 5% calf serum. P86, on the other hand, grew in all combinations except transferrin alone, PDGF alone and EGF with FGF. P86 also grows significantly better in media containing all supplemented factors than the parental NIH3T3 line. Thus it appears that the difference between P86 and NIH3T3 is a generally better growth rate in defined media, rather than a loss of response and requirement for PDGF or FGF.

Table 5. Comparison of growth of P86 and NIH3T3 in defined media supplemented with single growth factors and pairs of growth factors (Section 6.5).

Supplements added × transferrin	NIH3T3 Cell density	Doublings	P86 Cell density	Doublings
No supplements	$<1.0 \times 10^4$	<1.0	$<1.0 \times 10^4$	<1.0
Insulin	$<1.0 \times 10^4$	<1.0	3.2×10^4	1.7
EGF	$<1.0 \times 10^4$	<1.0	4.9×10^4	2.3
PDGF	$<1.0 \times 10^4$	<1.0	1.7×10^4	<1.0
FGF	1.3×10^4	<1.0	3.6×10^4	1.8
Insulin, EGF	$<1.0 \times 10^4$	<1.0	7.4×10^4	2.9
Insulin, PDGF	1.2×10^4	<1.0	4.8×10^4	2.3
Insulin, FGF	5.5×10^4	2.5	7.4×10^4	2.9
EGF, PDGF	$<1.0 \times 10^4$	<1.0	5.4×10^4	2.4
EGF, FGF	$<1.0 \times 10^4$	<1.0	1.8×10^4	<1.0
Insulin, EGF, PDGF	1.8×10^4	<1.0	1.1×10^5	3.4
All four	4.3×10^4	2.1	1.9×10^5	4.2
5% calf serum	1.2×10^6	7.0	9.6×10^5	6.6

8. ACKNOWLEDGEMENTS

This work was supported by NIH grant CA 46547 and by American Cancer Society grant MV-251.

9. REFERENCES

1. Chiang, Lin-Chiang, Silnutzer, J., Pipas, J. M. and Barnes, D. W. (1984) *Methods for Serum-Free Culture of Epithelial and Fibroblastic Cells.* Alan R. Liss, Inc, New York, p. 265.
2. Barnes, D. and Sato, G. (1980) *Anal. Biochem., 102*, 255.
3. Barnes, D. and Sato, G. (1980) *Cell, 22*, 649.
4. Barnes, D. (1987) *Biotechniques, 5*, 534.
5. Chiang, Lin-Chiang, Silnutzer, J., Pipas, J. M. and Barnes, D. W. (1985) *In Vitro Cell. Dev. Biol., 21*, 707.
6. Greenwood, D. C., Pogue-Geile, K., Srinivasan, A., McGoogan, S., Meisler, A. I. and Pipas, J. M. (1988) In *IBD: Current Status and Future Approach.* MacDermott, R. P. (ed.), Elsevier Science Pub. B.V., Amsterdam, p. 493.
7. Loo, D. T., Fuquay, J. I., Rawson, C. L. and Barnes, D. W. (1987) *Science, 236*, 200.
8. Nistar, M., Hammacher, A., Mellström, K., Siegbahn, A., Ronnstrand, L., Watermark, B. and Heldin, C. (1988) *Cell, 52*, 791.
9. Stoobant, and Waterfield, (1984) *EMBO J., 2*, 2963.
10. Jainchill, J. L., Aaronson, S. A. and Todaro, G. J. (1969) *J. Virol., 4*, 549.
11. Todaro, G. J. and Green, H. (1963) *J. Cell Biol., 17*, 299.
12. Reznikoff, C. A., Brankow, D. W. and Heidelberger, C. (1973) *Cancer Res., 33*, 3231.
13. Southern, P. J. and Berg, P. (1982) *J. Mol. Appl. Genet., 1*, 327.
14. Graham, F. L. and van Der Eb, A. J. (1973) *Virology, 52*, 456.
15. Frost, E. and Williams, J. (1978) *Virology, 91*, 39–50.
16. Xi Zhan and Goldfarb, M. (1986) *Mol. Cell. Biol., 6*, 3541.
17. McKeehan, W. L. and Ham, R. G. (1976) *J. Cell Biol., 71*, 727.
18. Pardee, A. B., Chevington, P. V. and Medran, E. E. (1981) In *Methods for Serum-Free Culture of Epithelial and Fibroblastic Cells.* Alan R. Liss Inc., New York, p. 157.
19. Peden, K. W. C., Pipas, J. M., Pearson-White, S. and Nathans, D. (1980) *Science, 209*, 1392.
20. Perucho, M., Hanahan, D. and Wigler, M. (1980) *Cell, 22*, 309.

CHAPTER 4

Growth of human leukaemia cells *in vitro*

BEVERLY J. LANGE

1. INTRODUCTION

There are three practical reasons to grow leukaemic cells *in vitro*: (i) to characterize and enumerate the leukaemic progenitors; (ii) to define the molecules that
regulate leukaemic proliferation and differentiation; and (iii) to develop culture
and sensitivity assays for leukaemic cells and chemotherapeutic agents analagous
to those for bacterial sensitivity to antibiotics. Unfortunately, these purposes are
thwarted by the lack of a consistent cell culture system to grow all human
leukaemias.

In vivo leukaemic cells enjoy a selective growth advantage over normal
haematopoietic cells—at least in those individuals who have the disease we
recognize as leukaemia. *In vitro* the opposite occurs: leukaemic cells are so
difficult to grow and maintain that failure to form any colonies *in vitro* is a
characteristic by which some myeloid leukaemias are classified (1). *In vitro* it is the
normal cells that have the selective growth advantage. This advantage is
demonstrated by failure to establish long-term Dexter-type suspension cultures of
cells of patients with Philadelphia chromosome (Ph^1) positive chronic myeloid
leukaemia. Over a period of weeks or months the number of cells with the Ph^1
karyotype decreases until it is undetectable and the number of normal mitoses
increases to 100% (2). The relative inability to survive *in vitro* under these
conditions is now being exploited to purge human marrow of residual leukaemic
cells in order to return it to the patient some days later in the form of a
leukaemia-free autologous marrow transplant (3).

Despite the propensity for normal cells to outgrow leukaemic cells *in vitro* or for
the leukaemic cells to die, several methods for growing leukaemic cells in
short-term culture have been developed. The method most often used involves
leukaemic colony-formation in soft agar or methylcellulose (4). Leukaemic
colony-formation assays are derived from systems for colony-formation of normal
haematopoietic progenitors in semi-solid media (4–6).

The second method is growth in suspension culture. Most leukaemic cells
survive for only a few days in culture media with foetal calf serum but even this
short life *in vitro* suffices for some investigations (6). Occasionally, the cells in
suspension survive for weeks, and rarely they may form established cell lines
(6–9). Some established cell lines survive *in vitro* because of the presence of
transforming viruses such as the Epstein–Barr virus (EBV) or human T-cell
leukaemia virus I (HTLV-I) or II (HTLV-II) (10–12), but others are free of

61

detectable virus. Probably only a portion of the whole leukaemic population is represented in these leukaemic colonies, suspension cultures and lines. Finally, a third method to support growth of leukaemic cells is the xenografting of human cells into privileged sites such as the anterior chamber of the eye of the nude mouse, a method that does not lend itself to a book entitled *A Practical Approach* (13).

It is now apparent that growth of haematopoietic cells and of leukaemic cells *in vitro* and *in vivo* is the result of complex interactions between colony-stimulating factors, growth factors and growth factor receptors. These factors may be humeral factors present *in vivo* in plasma, they may be autocrine factors generated by the leukaemic cells themselves, and they may be paracrine factors that are dependent on cell-to-cell interactions in the haematopoietic microenvironment. Recently, recognition, purification and production of these factors in recombinant systems has allowed more precise definitions of conditions for growth of leukaemic cells; ultimately they will provide reproducible leukaemic colony-forming and suspension cultures and will explain the relationship of the subpopulations of leukaemic cells to the putative leukaemic progenitor.

This chapter describes the methods for leukaemic colony-formation and for growth in short- and long-term suspension culture. For a more detailed description of colony-formation in normal haematopoietic cells the reader is referred to the comprehensive manual by Metcalf (4).

2. HUMAN LEUKAEMIC COLONY-FORMATION *IN VITRO*

A colony is arbitrarily defined as a group of 20 to 40 or 50 cells that has originated from a single cell over 7–14 days *in vitro* (1). The techniques for growing leukaemic colonies were first adopted from those of Pike and Robinson for normal haematopoietic colony-formating progenitors and then modified to the particular requirements of leukaemic cells. Most studies of leukaemic colony-formation deal with acute non-lymphoblastic leukaemic (ANLL) cells, the prototype leukaemic colony assay; thus, most of this chapter will refer to techniques developed for ANLL. With current technology has come the ability to grow 80–90% of ANLL cells in colony assays. Systems in acute lymphoblastic leukaemia and chronic myeloid leukaemia for colony-formation are described in Sections 2.4 and 2.5.

2.1 Obtaining human leukaemic cells

Leukaemic cells can be obtained from peripheral blood, bone marrow aspirate or biopsy or rarely from a chloroma or spleen. Peripheral blood offers the advantage that multiple specimens can be withdrawn without much discomfort to the patient. The disadvantages are (i) peripheral blood may contain many non-leukaemic cells, especially T-lymphocytes and (ii) the relationship of the circulating leukaemic population to the marrow population is not well understood and is probably variable from one patient to another.

Table 1 summarizes the procedure for obtaining human leukaemic marrow. The use of human leukaemic blood or marrow for research usually requires approval

Table 1. Procedure for obtaining human leukaemic marrow for culture.

Materials

Previous Institutional Review Board approval for leukaemic marrow
Two syringes $\geqq 10$ ml in vol.
Marrow aspiration needle
Sterile 18-gauge needle
Preservative-free heparin
Alcohol, betadine
Local or general anaesthetic

Method

1. Obtain consent.
2. Cleanse skin over iliac crest or sternum with betadine and alcohol.
3. Anaesthetize skin and periosteum.
4. Draw up 100–500 U pan heparin into syringe (100 U/1 ml marrow).
5. Insert aspiration needle.
6. Withdraw 0.2–5.0 ml into dry syringe for diagnostic studies.[a]
7. Change syringe.
8. Withdraw 1–5 ml into heparinized syringe.
9. Rotate syringe to mix marrow with heparin.

[a] If no diagnostic studies are needed, steps 6 and 7 can be omitted.

of the Institutional Review Board (IRB) and should entail some assent or consent on the part of the patient. Many IRBs do not require a separate written consent for cells if the procedure to obtain the cells is otherwise medically necessary for the evaluation of the patient. Usually no consent is required for use of autopsy material; however, to be viable, the autopsy specimens must be obtained within several hours of death.

Marrow aspiration, while it is not an especially painful procedure when the marrow is normal, is often extremely painful when the marrow is full of leukaemia. For this reason the number of aspirations performed on patients with leukaemia is limited and an attempt should be made to obtain as much information from a single aspirate as possible. Thus, in an untreated patient it is often necessary to obtain a 0.2–1.0 ml in a dry syringe for smears for diagnostic morphologic, histochemical and cytogenetic studies. The use of preservative-free heparin interferes with the quality of these studies. Then a second, heparinized syringe is used to aspirate liquid marrow. Because the marrow is hypercellular and may be packed with blasts, a 30-, 50- or 100-ml syringe may be needed. Occasionally no marrow can be aspirated. In this case a marrow biopsy is obtained with a Jam-Shihidi needle. Cells can then be scraped from the biopsy with a scalpel, teased out with a needle or gently washed out with a syringe and 21-gauge needle. Often some cells are injured, but a number sufficient for investigation remain.

2.2 Preparation of leukaemic cells for culture

The procedure for preparing leukaemic cells for culture entails removal of the acidic heparin which may harm the cells, removal of the patient's serum which may inhibit growth, removal of erythrocytes which make it difficult to see the leukaemic cells, removal of cells which may inhibit leukaemic growth such as granulocytes, and sometimes removal of those which may stimulate growth, such as T-lymphocytes and monocytes, or which may themselves form colonies difficult to distinguish from leukaemic colonies.

2.2.1 *Density gradient separation of mononuclear cells*

The most common method of cell separation involves a density gradient, Ficoll–Hypaque (Ficoll–Isapaque, Pharmacia, Uppsala), which separates cells with a specific gravity of greater than 1.077 from those that are less dense (*Table 2*) (14). Before placing cells on a gradient, dilute marrow or blood with a physiologic phosphate-buffered salt solution; overlay the diluted specimen on the gradient and centrifuge as described (*Table 2*). The bouyant cells at the gradient–specimen interface include lymphocytes, monocytes, neutrophil precursors from blast to myelocyte stage, proerythroblasts and erythroblasts, some platelets and leukaemic cells. Erythrocytes, granulocytes and dead cells are in the pellet. Usually 90% of the mononuclear cells are recovered in this process. The cells require two additional washes to remove the hypertonic Ficoll–Hypaque; the second wash must include at least 1% serum to prevent the cells from adhering to one another. About 5–10% of cells are lost with each wash. After washing, resuspend the cells in medium and serum and count. Cell yield is highly variable from approximately 5×10^6 cells per ml of leukaemic marrow to an average of approximately 20×10^6/ml to a maximum 40–50×10^6/ml. Peripheral blood yields 10^6 mononuclear cells for each 2×10^3/ml.

Table 2. Preparation of low density human leukaemia cells for culture.

Materials

Sterile laminar air flow hood
Heparinized marrow or blood
Ca^{2+}, Mg^{2+}-free PBS, pH 7.2 or similar phosphate-buffered balanced salt solution
15-ml pointed centrifuge tubes
Ficoll–Hypaque
Tissue culture medium (Iscove's, McCoy 5A, MEM-α)
Serum (usually FCS)

Method

1. Dilute marrow 1 : 4 with PBS or dilute peripheral blood 1 : 2 with PBS.
2. Place 5 ml of Ficoll–Hypaque in centrifuge tube.
3. Overlay 8 ml of diluted marrow or blood on Ficoll gradient.
4. Centrifuge cells for 20 min at room temperature.
5. Remove mononuclear cell interface.
6. Resuspend in 5 vols of PBS with 1/ FCS in centrifuge tube.
7. Centrifuge at 1200 *g* for 10 min.
8. Resuspend in 5 vols PBS with 1% FCS.
9. Count cells.

Swart *et al.* have described a discontinuous albumin gradient, with specific gravity from 1.056 to 1.080. This gradient further separates subpopulations of light-density mononuclear cells; the population of blasts with a density of greater than or equal to 1.062 has the greatest capacity to form colonies (15). The technique has value for investigations aimed at isolating the ultimate leukaemic progenitor cell and characterizing the leukaemic subpopulations, but is time-consuming and cumbersome and is not generally used for simple studies quantitating colonies and clusters and response of the leukaemic population to various growth factors.

2.2.2 *Removal of adherent cells*

After density-gradient separation the only adherent cells remaining at the interface are monocytes and macrophages. These cells are the source of colony-stimulating factors for normal and leukaemic myeloid colony-forming cells as well as for lymphocytes, which both inhibit and stimulate colony-formation. The necessity for removal of these cells depends on the purpose of the colony assay. If the purpose is to measure the isolated effect of a purified growth factor on the leukaemic cells, then adherent cells need to be removed; if it is to examine the global effect of the factor on the leukaemic cells as well as other cells in the haematopoietic microenvironment, the adherent cells remain. If the purpose is to examine the effect of a drug, or to compare growth patterns of leukaemia from a variety of patients, generally, they are not removed. All mononuclear cells are necessary in a feeder layer for acute myeloid leukaemia colony-formation. The procedure for adherent cell removal is given in *Table 3*.

2.2.3 *Removal of sheep erythrocyte rosetting cells (T-lymphocytes)*

The same caveats and rationale for removing contaminating adherent cells pertain to removing sheep erythrocyte rosetting (E-rosette[+]) T-lymphocytes. Furthermore, a number of colony-formation assay systems use phytohaemagglutinin (PHA) directly or phytohaemagglutinin–lymphocyte-conditioned-medium (PHA-LCM) to stimulate colony-formation. PHA is a potent stimulator of T-lymphocyte proliferation, blastogenesis and agglutination. Thus, if there are many contaminating T-lymphocytes, distinguishing aggregates of leukaemic blasts from aggregates of stimulated T-lymphocytes is difficult. Hence, the need to remove a substantial number of T-lymphocytes (16). The sheep erythrocyte

Table 3. Separation of adherent cells from leukaemic cells.

1. Resuspend washed mononuclear cells in medium with 10% FCS at 1×10^7/ml.
2. Place in plastic tissue culture flask at 10–20% vol. of flask.
3. Place flask horizontally in 37°C incubator for 3 h.
4. Gently remove the supernatant fluid and cells.
5. Repeat once if needed.
6. Count cells.

rosetting technique outlined in *Table 4* removes over 90% of peripheral blood T-lymphocytes. It also removes leukaemic lymphoblasts of T-cell origin that express the sheep erythrocyte receptor and the rare myeloid leukaemic blasts that aberrantly express the receptor. The T-lymphocytes in the pellet can be recovered by deftly lysing the attached erythrocytes in a hypotonic solution such as distilled water or Tris-buffered NH_4Cl.

The use of a pan-T monoclonal antibody and complement remove the contaminating T-cells more efficiently. However, many leukaemic cells are exquisitely sensitive to most forms of complement and they may be lysed in the process.

Table 4. Separation of leukaemic cells from peripheral blood T-lymphocytes.

Materials

Washed and gradient-separated leukaemic blood or marrow at 2×10^6/ml in PBS, 1% FCS
Neuraminidase (CALBIOCHEM)
Sterile sheep erythrocytes (SRBC) 50% (v/v) in Alsever's solution
Ficoll–Hypaque gradient
PBS
Sterile distilled H_2O

Method

A. Preparation of neuraminidase-treated SRBC (NSE-SRBC)
   1. Dilute 1 ml of SRBC in 9 ml of PBS.
   2. Add 1 ml of neuraminidase to 9 ml of SRBC-PBS.
   3. Mix by inverting.
   4. Incubate for 45 min in a 37°C water bath, inverting twice.
   5. Wash twice in PBS (v/v, 1/5).
   6. Wash twice in FCS (v/v, 1/1).
   7. Resuspend in 9 ml of FCS.

B. Formation of SRBC rosettes (E-rosettes).
   1. Add 1 ml of NSE-SRBC to 3 ml of cell suspension.
   2. Mix.
   3. Incubate in a 37°C H_2O bath for 10 min.
   4. Centrifuge at 1500 r.p.m. at 4°C for 5 min.
   5. Place tube on ice for 30 min.
   6. Gently rock to resuspend.
   7. Underlay 6 ml of Ficoll-hypaque.
   8. Centrifuge at 2000 r.p.m. at 4°C for 12 min.
   9. Collect non-rosetting cells at the Ficoll interface.
  10. Repeat once.
  11. Wash non-rosetting cells in PBS and 1% FCS.
  12. Resuspend in medium and 10–20% FCS.

C. Recovery of T-lymphocytes
   1. Remove cell gradient and supernatant fluid, leaving pellet.
   2. Add 1 ml of H_2O.
   3. Mix rapidly by pipetting with Pasteur pipette to lyse erythrocytes (10 sec).
   4. Add 4 ml of medium 10% FCS.
   5. Wash once.

Complement must be pre-adsorbed with human AB^+ erythrocytes as well as the blasts themselves at 4°C. A pan T-cell antibody and a cell sorter can also be used to eliminate T-lymphocytes but are less sensitive than the rosetting.

2.2.4 *Cell concentration*

The efficiency of colony and cluster formation varies enormously from patient to patient. If cell numbers are sufficient, final cell concentrations of 10^4, 5×10^4, 1×10^5 and 2×10^5/ml should be assayed in triplicate or quadruplicate. If cell numbers are limited a small aliquot can be tested and the remainder of cells frozen in liquid nitrogen in 50% foetal calf serum (FCS) and 10% dimethylsulphoxide. The number of colonies is difficult to count when it is over 100 or when there are hundreds of clusters; furthermore, a large number of colonies may deplete metabolites. About 30 colonies per dish allows one to count the colony number accurately and withdraw isolated colonies. Finally, the relationship of leukaemic colonies to number of cells plated may not be linear, suggesting that local intercellular environmental factors may contribute to colony-formation or, less commonly, to colony-inhibition.

2.3 **Procedure for colony-formation in ANLL**

2.3.1 *Materials*

Table 5 lists the materials and equipment needed for growth of leukaemic colonies *in vitro* and *Tables 6* and *7* describe the details of an actual experiment in agar (*Table 6*) or methylcellulose (*Table 7*) involving triplicate or quadruplicate 1-ml samples.

(i) *Medium.* Iscove's media, McCoys 5A, Dulbecco's modified Eagles' medium, and minimal essential media-α are best suited to normal haematopoietic and leukaemic colony-formation. Glutamine should be added at the time of use. Penicillin and streptomycin are not needed if care is taken with techniques; antibiotics may inhibit certain individual's cells and contamination in humidified incubators is often fungal. Appropriate positive and negative controls must be included within each experiment.

(ii) *Serum.* Foetal calf serum is usually used in concentrations of 10–20%. Before use, pretest lots of serum for their ability to support normal haematopoietic growth or growth of leukaemic cells known to form colonies *in vitro*. Most vendors are willing to send 100-ml samples for pretesting. It is wise to order a large supply of FCS from a pretested lot that supports growth and to store the supply at −20°C. Human serum or plasma support the growth of some leukaemic cells.

(iii) *Agar.* Agar can be adequately sterilized by boiling for 2 min; longer boiling will concentrate the agar; 3% agar will not gel at 37°C but gels quickly at room temperature (*Table 5*). For this reason the agar and cells must be kept at 37°C; a higher temperature will kill the cells. Warmed pipettes help to slow the gelling.

Table 5. Human leukaemic colony-formation in soft agar.

Materials

Density-separated leukaemic cells
Iscove's medium
Pretested FCS
Source of CSF (see *Table 7*)
35-mm Petri dishes (Lux, Flow Laboratories)
Double distilled deionized H_2O
Bacto–Agar (Difco) or Noble Agar (Difco)
5- or 10-ml pipettes
10-ml tubes
Laminar air flow hood
Humidified CO_2 incubator
37°C water bath

Methods (For quadruplicate 1-ml cultures)

1. Label Petri dishes with waterproof marker.
2. Resuspend cells at 62.5×10^5–1.25×10^6/ml in medium with 25% FCS.
3. Hold cells at 37°C.
4. Mix agar with H_2O 3% (v/w).
5. Boil for 2 min.
6. Hold at 37°C.
7. Add 0.5 ml of CSF to 4 ml of cell suspension.
8. Using a warm pipette, add 0.5 ml of agar to the cell–CSF mixture.
9. Quickly mix by pipetting.
10. Place 1 ml in a Petri dish.
11. Swirl to mix.
12. Place 1 ml in the next Petri dish, swirl, etc.
13. Discard the last 1 ml.
14. Allow to gel at room temperature (10–20 min).
15. Cover.
16. Incubate for 7–14 days at 37°C in 5% CO_2.

Add the agar to the cell mixture and rapidly pipette up and down five times to mix; then add 1-ml aliquots quickly to the Petri dishes. Mix each dish immediately by swirling so the agar coats the bottom and sides evenly. The agar will gel in 10–20 min. Do not cover the dishes until gelling is complete to avoid condensation on the lid: condensation will kill the cells.

(iv) *Methylcellulose.* Methylcellulose at 0.8–1.0% can also support leukaemic colony-formation. Large volumes of methylcellulose can be made and frozen (*Table 6*). Some methylcellulose is toxic to human cells; 4000 centipoise (Fischer) generally works well. Methylcellulose is a thick liquid that allows movement of cells. Agar offers the advantage that it is a true gel and colonies are fixed in it. Colonies can be plucked more easily from agar but they are difficult to disperse thereafter. Colonies can be removed from methylcellulose using a Pasteur pipette drawn to a fine taper and then dispersed using forced air from a jet. The dispersed

Table 6. Human leukaemic colony-formation in methylcellulose.

Materials

Leukaemic cells
Iscove's medium 2× at 4°C
Iscove's medium 1×
FCS
Methylcellulose 4000 centipoise (Fischer)
Sterile, double distilled deionized H_2O
2-litre flask
Magnetic stirrer
10-ml tubes
Autoclave
5-ml syringes
>3″ large long bore >16-gauge needles such as Jam-Shihidi needles

Method

A. 2.7% Methylcellulose stock
   1. Place 14 g of methylcellulose in a 2-litre flask; autoclave.
   2. Add 250 ml of boiling H_2O.
   3. Stir with sterile magnetic stirrer.
   4. Cool to 40°C.
   5. Add 250 ml 2× Iscove's medium at 4°C.
   6. Agitate vigorously for 24–48 h at 4°C.
   7. When clear and free of lumps, freeze at 4°C in 50 ml aliquots.

B. Cultures (For triplicate 1-ml cultures)
   1. Resuspend cells in Iscove's medium at 2×10^6/ml.
   2. For 5 ml
0.5 ml of cells	(10%)
1.0 ml of FCS	(30%)
0.5 ml of CSF	(10%)
0.8 ml of medium	(2.0%)
1.7 ml of methylcellulose	(0.9%)
   3. Cap tube.
   4. Mix by vortexing or with repeated pipetting with syringe and needle.
   5. Dispense 1 ml in each Petri dish.
   6. Disperse by coating sides and bottom of dish with slow swirling motion.
   7. Cover.
   8. Incubate in humidity, at 37°C, in 5% humidified CO_2 for 7–14 days.

cells can be fixed, stained and individual cells identified. On the other hand, the whole agar plate can be fixed, stained and examined (see Section 2.4).

(v) *Colony-stimulating activity and colony-stimulating factors.* To form colonies *in vitro* normal haematopoietic cells and virtually all myeloid leukaemic cells require an exogenous source of colony-stimulating activity (CSA). The only exceptions are rare leukaemias which have autonomous production of granulo-cyte–macrophage colony-stimulating factor (GM-CSF) (17) and some established leukaemic cell lives that have lost their absolute CSA dependence (9–18). As in

Table 7. Sources of colony-stimulating activity (CSA).

Leukocyte feeder layer (5, 22)

Materials

   Peripheral blood mononuclear leukocytes (PBL) (*Table 2*)
   Noble Agar 1% (w/v) (Difco) (*Table 5*)
   2× Iscove's medium
   FCS
   37°C water bath
   35-mm Petri dishes (Lux)

Method

   For 10 1-ml Petri dishes:
   1. Resuspend PBL at 2×10^6/ml in 2× Iscove's medium with 20% FCS.
   2. To 5 ml PBL add 5 ml agar; mix; distribute in 1-ml aliquots in Petri dishes.
   3. Cool for 10 min; cover.
   4. Incubate at 37°C in 5% humidified CO_2.

Phytohaemagglutinin-stimulated leukocyte conditioned medium (PHA-LCM) (23, 24)

Materials

   PBL
   FCS
   Medium RPMI 1640
   Reagent Grade PHA (9 mg/100 ml) (Wellcome)

Method

   1. Resuspend PBL at 1×10^6/ml in RPMI 1640 2% FCS.
   2. Add PHA 1% (v/v); incubate at 37°C in 10% CO_2 for 3–4 days.
   3. Collect and filter supernatant fluid; store at 4°C.

Mixed lymphocyte reaction—leukocyte conditioned medium (MLC-LCM) (21, 23)

Materials

   PBL from three unrelated donors
   FCS
   Medium RPMI 1640

Method

   1. Mix equal volumes of PBL from each donor at 1×10^6/ml in medium 2% FCS.
   2. Incubate at 37°C in 10% CO_2 for 7 days.
   3. Collect and filter supernatant fluid; store at 4°C.

Malignant human cell-line conditioned medium

   Giant cell tumour CM (Gibco) (25)
   MO-T (HTLV-II) leukaemic CM (American Tissue Culture Association) (12)
   5637 Bladder Carcinoma (J. Fogh, Sloan-Kettering Institute, Rye, NY) (26)

the case of normal haematopoietic cells CSA induces proliferation and is essential for leukaemic cell survival, differentiation, and lineage-specific functions. Response of leukaemic cells to CSAs varies greatly from patient to patient.

The notation CSA is used when a mixture of poorly defined colony-stimulating factors is used; CSF refers to specific factors. Those CSFs which clearly have activity for ANLL colony-formation are colony-stimulating factor GM-CSF and interleukin-3 (IL3). Granulocyte–CSF (G-CSF), originally thought to induce differentiation, has been shown to augment GM-CSF effects in normal haematopoietic cells; the effects of G-CSF appear to be additive to those of GM-CSF in leukaemic haematopoiesis as well (19). (Leukaemic proliferation may also be suppressed by CSF, especially G-CSF (20) and by interferons.) The role of interleukin-2 (IL2) in ANLL is unclear. Human erythroleukaemia may grow in the absence of GM-CSF, but responds optimally to a combination of erythropoietin and GM-CSF (21).

Sources of CSA and CSF are listed in *Tables 7* and *8*. The leukocyte feeder in *Table 7* consists of 10^6 peripheral blood mononuclear cells in 1 ml of 0.5% agar in medium and FCS (22). Monocytes, T and B lymphocytes are necessary to ANLL proliferation. Leukocyte feeder layers vary from donor to donor and may lose activity or become contaminated after storage in the CO_2 incubator after 1 week. Some of the cells in the feeder layer may themselves form compact colonies which can confound counting. Feeder layer colony-formation can be prevented by irradiating the underlayer with 2500 rads. Either methylcellulose or soft agar (0.3%) can be overlaid on the leukocyte feeder layer.

The limitations of the leukocyte feeder layer have led to the use of liquid CSA. Common sources of CSA are PHA-LCM, human tumour cell line conditioned medium and now recombinant conditioned medium. PHA-LCM suffers from variability among donors, but those donors whose cells are a potent source of PHA-LCM usually remain relatively potent upon repeated collection. An alternative to PHA-LCM is a mixed lymphocyte reaction (MLR-LCM). The peripheral blood mononuclear cells of at least three donors are mixed at 10^6/ml for 7 days and the media is harvested as above for PHA-LCM (*Table 7*) (23, 24). This MLR-LCM method has the advantage of being free of PHA which may stimulate contaminating T-lymphocytes among the leukaemic cells. PHA-LCM and MLC-LCM must be filtered to eliminate contaminating donor cells.

Human tumour cell lines are reproducible sources of admixtures GM-CSF and other haematopoietic growth factors. Supernatant fluid from the giant cell tumour (GCT) line available from Gibco has had consistent activity from lot to lot (25). The Mo-T cell line can be obtained from the American Tissue Culture Association. The bladder carcinoma cell line 5637 can be purchased from Dr J. Fogh, Sloan and Kettering Institute for Cancer Research, Rye, NY. Maintenance of the lines themselves provides a relatively inexpensive reproducible source of GM-CSF, provided culture and harvesting conditions are not modified over time (26).

Methods for purification of GM-CSF (27), and IL3 (28) are available (*Table 8*). Use of purified growth factors allows more precise definition of the growth conditions for leukaemic cells. Recombinant G-CSF, GM-CSF, and IL3 can be

Table 8. Sources of colony-stimulating factors (CSF).

Purified CSF
    Granulocyte–monocyte CSF (GM-CSF) (27)
      (Prepared from Mo-T line or 5637 cell-line grown in serum-free medium)

    Interleukin-3 (IL3) (28)
      (Prepared from PHA, TPA stimulated leukocytes or 5637 cell-line)

Recombinant CSF
    GM-CSF (Genetics Institute, Cambridge, Massachusetts) (30)
      (GM-CSF Chinese hamster ovary (CHO) cells)

    G-CSF (Genetics Institute, Cambridge, Massachusetts) (31)
      (G-CSF transfected monkey COS-1 cells)

    IL3 (Genetics Institute, Cambridge, Massachusetts) (29)
      (IL3 transfected monkey COS-1 cells)

    IL2 (Biogen Institute, Geneva) (37)
      (IL2 transfected *E.coli*)

---

obtained from the Genetics Institute, Cambridge, Massachusetts. Recombinant CSFs appear to have the same activity as the purified factors (29–32).

The media in *Table 7* require 10% CSF (v/v), a volume which is appropriate for PHA-LCM and most tumour cells line CM. For recombinant CM or purified CSFs smaller volumes are used based on a biologic assays of activity. Often they are added to the Petri dish directly to avoid the 20% waste that is inherent in the agar or methylcellulose system (*Tables 5* and *6*).

2.3.2 *Incubation*

Proper maintenance of the incubator is critical to colony-formation. After plating the cells, place the dishes in a clean humidified incubator at 37°C with CO_2. Check the CO_2 supply, the CO_2 concentration, the temperature and humidity twice daily: colony-forming cells do not tolerate abrupt changes in CO_2 or temperature. For the same reason resist removing the plates from the incubator except on days 7 and/or 14. If they must be removed only a small number should be examined at one time. Likewise, avoid excessive opening and closing of the incubator.

2.3.3 *Scoring the experiment*

Scoring the experiment consists of counting colonies and clusters in each dish, comparing the experimental cultures to positive and negative controls, measuring the variability among triplicate or quadruplicate samples, and determining the composition of colonies. The methods described result in cluster formation in 50–75% of cases of ANLL and colony-formation in about 25–40% of cases. Efficiency can be increased by using the leukocyte feeder layer in combination with preincubation of T-lymphocyte-depleted leukaemic blasts in PHA for 15 h in a liquid or methylcellulose overlayer (15).

(i) *Enumerating*. Metcalf emphasizes that the 'scoring of colonies is the responsibility of the investigator' (4). It is generally agreed that aggregates of 3–20 cells are a cluster. Some consider 20–40 or 50 cells a colony; others consider these large clusters; anything over 50 cells is a colony. Colonies and clusters are counted on days 7 and 10 or 14 using an inverted microscope or a dissecting microscope. To count, place a 35-ml Petri dish over a scored 50-mm dish and move the dish manually or with a stage. Focus up and down to detect colonies in each field at all depths. Pay particular attention to the preferential distribution of colonies at the edge of the dish. With experience it is possible to assess each aggregate in one or two seconds.

(ii) *Determining the composition of colonies*. Removal of isolated colonies requires practice and patience. Colonies are removed using a Pasteur pipette drawn to a fine point. Hold the plate in one hand; under direct vision, scoop the colony from the agar onto a slide coated with albumin or 2% FCS. Gently spread the colony with the pipette tip or a coverslip, fix with methanol and stain with May–Grünwald–Giemsa. Colonies in methylcellulose can be aspirated into the pipette and expelled [Section 2.3.1(iv)]. Always check the plate after colony removal to see if the colony has really been removed.

Entire agar cultures can be fixed in glutaraldehyde using the method of Salmon and Buick (33). One millilitre of 2.5% glutaraldehyde in phosphate-buffered saline (PBS) is added to the dish for 1 h. To obtain the culture, loosen the agar from the dish with a spatula, float it onto a water bath, then float it onto an albumin-coated slide, cover the agar with wet filter paper, and gently press and allow it to dry; remove the paper, stain colonies with May–Grünwald–Giemsa and cytochemical stains for peroxidase and specific and non-specific esterase. Usually the colonies derived from leukaemic blasts consist of a combination of morphologically identifiable blasts and dysplastic early myeloid or monocytic precursors. Auer rods may confirm the leukaemic origin of the cells. More rigorous proof of their leukaemic origin can be obtained by determining the karyotype of cells in very large colonies ($\geq$1000 cells) using the technique of Dube *et al.* (34) or by demonstrating clonality.

2.3.4 *Replating of leukaemic colonies*

Most colonies are comprised of cells with limited proliferative potential; clusters have even less proliferative potential than colonies. Proliferative potential can be demonstrated by secondary plating of the colonies. Leukaemic colonies can be replated either by pooling all cells in a dish or by dispersing and replating single colonies using precisely the same conditions as for primary colonies. The efficiency of colony-formation declines with the second passage. Subculturing beyond four replatings has not been successful (35).

The procedures described above have been used for normal haematopoietic cells and ANLL blasts. Those for chronic myeloid leukaemia (CML) and acute lymphoblastic leukaemia (ALL) are described below.

2.4 Procedure for colony-formation in CML (chronic myeloid leukaemia)

Peripheral blood cells and marrow from untreated patients with CML in chronic phase proliferative luxuriously in the conditions described above: marrow often yields 100-fold more granulocyte–monocyte and eosinophil colonies than normal marrow; peripheral blood, greater than or equal to 1000-fold more colonies than normal blood. Thus, in the early phase, 10^3 or 10^4 cells can be plated in each dish (Section 2.2.4). Procedures are then precisely as described for ANLL (Section 2.3). As CML progresses, colony-formation decreases and may become like that in ANLL.

2.5 Procedure for colony-formation in ALL (acute lymphoblastic leukaemia)

Systems for blast cell colony-formation in ALL are more complex than those for ANLL or CML. They resemble systems for normal T-lymphocyte colonies in that they require (i) an inducer of interleukin-2 (IL2) receptors, and (ii) a source of IL2 itself. PHA or tetradecanoylphorbol-13-acetate (TPA) are efficient inducers of IL2 receptors. Crude IL2 can be obtained from PHA T-lymphocyte-conditioned medium; purified IL2 can be obtained by gel filtration or by recombinant DNA technology (*Table 8*).

2.5.1 ALL of B-cell lineage

Table 9 lists two systems for colony-formation in B-lineage ALL (36, 37). Either system yields colonies of undifferentiated pre-B-lymphocytes in over 80% of cases. The single layer methylcellulose method is complicated by its requirement for PHA-LCM obtained from patients with haemachromatosis and for a specialized CO_2, O_2, N_2 incubator. Also, the leukaemic blasts must be mixed directly with feeder cells—if the feeder cells are in a separate agar layer they fail to stimulate colony-formation. The feeder cells and their debris make scoring more difficult.

The two-layer system has the advantage of using feeder cells from ostensibly normal human donors; feeder cells are sequestered in relatively firm agar (0.5%) and can be easily distinguished from leukaemic colonies. The medium used is rather complex but the added ingredients can be stocked and prepared in advance. This method also requires meticulous depletion of contaminating T-lymphocytes. Touw *et al.* (37) demonstrate that IL2 is a critical growth factor in this system and that blasts respond to IL2 in concentrations of 25–500 U/ml in contrast to normal cells which have a peak response to IL2 of 25 U/ml. Normal lymphocytes, however, do not require a feeder layer. The feeder layer is also essential for B-lineage blast proliferation. The feeder layer may be responding to IL2, to PHA or TPA, to the foreign blasts or to all of these factors. The interaction of these factors is not yet clear.

More recently, the effect of two B-cell growth factors (BCGFs) have been shown to stimulate tritiated thymidine incorporation in pre-B ALL blasts and to stimulate B-lineage ALL colony-formation (36). One is a 60000 dalton high-molecular-weight factor (HMW) purified from a T-cell ALL line and the other, a 12000 dalton low-molecular-weight factor (LMW) purified from PHA-LCM. The

Table 9. Colony-formulation in B-lineage acute lymphoblastic leukaemia.

A. *Single layer methylcellulose culture* (28)

Materials

   T-depleted leukaemic cells (*Tables 2* and *4*), 4×10^6/ml in a α med 20% FCS
   Normal PBL (adherent cells depleted) (*Table 3*) leukocyte feeder cells (*Tables 2* and *7*)
   PHA-LCM from donor with haemochromatosis (*Table 7*)
   Caesium radiation source or Mitomycin C
   Incubator: specialized modulated incubator for $O_2/CO_2/N_2$ (Billups-Rosenberg, del Mar, CA)
   α medium
   FCS
   2.7% methylcellulose (*Table 6*)

Method

   1. Incubate feeder cells at 1×10^6 in α medium, 10% FCS at 37°C for 2 h.
   2. Cool to 0°C on ice; irradiate with 2500 rads or treat with mitomycin-C 50 μg/ml at 37°C for 30 min and wash twice.
   3. For 5 ml (four quadruplicate dishes), mix the following:
        2.5 ml of leukaemic cell suspension
        0.5 ml of feeder cells
        0.5 ml of PHA-LCM
        1.5 ml of methylcellulose.
   4. Proceed to prepare dishes as in *Table 6*; incubate dishes for 7 days in 37°C incubation with 5% O_2, 5% CO_2, nitrogen. Cultures can also be prepared in 0.1-ml aliquots in a plated in 96-well microtiter plates (Linbro, Titertek, Hamden, CT); plates must be sealed lightly with tape.

B. *Double layer leukocyte feeder culture* (29)

Materials

   T-depleted leukaemic cells (*Tables 2* and *4*)
   Leukocyte feeder layer (*Table 7*)
   PHA
   TPA (Sigma)—10 ng/ml
   Dulbecco's modified Eagle's medium
     6.7% horse serum, 6.7% FCS
     6.7% trypticase soy broth and 10% BSA
   Source of IL2 (PHA-LCM; purified IL2; rIL2) (*Table 7*)
     (25–500 U/ml final concentration)

Method

   For each Petri dish:
   Place $0.2–1.0 \times 10^5$ leukaemic cells in 0.4 ml of MEM supplemented as above with 0.01 ml of PHA over the leukocyte feeder layer. Colonies are counted on day 7, 10 or 14.

HMW and LMW are often synergistic in their effects in colony-formation; LMW-BCGFs stimulated [³H]thymidine ([³H]Tdr) incorporation in 75% of cases; HMW-BSGF stimulated [³H]Tdr in 33% of cases. The fractionation analysis of response of pre-B ALL to these two factors is a first attempt at precise definition of factors involved in B-lymphocyte proliferation and differentiation and provides direct evidence for the existence of receptors to these two factors in pre-B ALL (38).

2.5.2 *ALL of T-cell lineage*

To form colonies, leukaemias phenotypically related to immature T-lymphocytes also require induction of IL2 receptors followed by IL2. As in the case of pre-B ALL, T-ALL blasts respond to either PHA or TPA in short-term (18-h) culture with demonstrable membrane IL2 receptors (39). Addition of IL2 to receptor-positive cells causes colony-formation in semi-solid media or thymidine uptake in suspension culture (39). In contrast to blasts from pre-B ALL, these cells do not require a feeder layer and thus do not seem to require the array of poorly defined factors apparently necessary for B-ALL proliferation *in vitro*. The procedure for colony-formation is outlined in *Table 9* except the feeder layer is omitted.

Few cases of T-ALL colony-formation have been studied in detail. From a sample of six cases, Touw *et al.* (39) demonstrate convincingly that IL2 receptor induction is needed in all cases: the need for exogenous IL2 itself varies from case to case. The acute lymphoblastic leukaemias of immature phenotype differ from the HTLV-I infected mature T-ALLs in their requirement for an exogenous factor that induces the IL2 receptor (see Section 4); in some cases they differ from malignant T-cell cutaneous lymphomas in requiring IL2 receptor induction and in variably demonstrating autocrine IL2 secretion (34). The role of IL2, TPA, antigen and other poorly characterized factors in T-ALL proliferation await definition.

3. PROCEDURE FOR GROWING LEUKAEMIC CELLS IN SUSPENSION CULTURE

3.1 **Short-term suspension culture**

Colony-formation assays are tedious, suffer from wide patient-to-patient variation and entail considerable difficulty in recovering clonogenic cells. Suspension cultures avoid these problems. When leukaemia cells are placed in a flask with medium and FCS, several things happen. Most often the number of cells increases 2-fold over a few days and then decreases progressively over 1 to 2 weeks until there are no viable cells (40). Sometimes cells form an adherent layer and the number of supernatant cells decline over several weeks. However, the adherent layer derived from leukaemic marrow does not usually support the growth of autologous leukaemic cells but may support normal cells under conditions used for a Dexter assay (2). The standard Dexter assay includes hydrocortisone, horse-serum, β-mercaptoethanol and all the other factors in standard colony assays (5). In these assays an adherent stromal layer of adipose cells, fibroblasts, macrophages and endothelial cells attaches to the floor of the flask. The adherent layer supports the outgrowth of mature leukocytes and colony-forming cells for weeks and months. The conditions of the Dexter assay inhibit for leukaemic cells and favour the outgrowth of normal cells. Leukaemic cells in suspension over a feeder layer usually show gradual dysmyelopoietic maturation. Occasionally they proliferate for weeks and then die either by lysis or terminal differentiation.

3.2 **Long-term suspension culture**

The methods for suspension culture and reculturing are given in *Table 10*. Addition of PHA-LCM or of haematopoietic growth factors to suspension cultures allows leukaemic cells to proliferate for longer than a few days (18, 41). in the presence of PHA-LCM, IL3 or GM-CSF, the blasts from the majority of patients with ANLL undergo as many as four doublings in 7 days but after 7 days most of the cells die. If colony-formation is assayed, the colony-forming ability of the supernatant cells remains consistent for the first week and then declines thereafter. There is no apparent recruitment of clonogenic cells under these conditions. In less than half the cases exponential growth persists for 1 to 2 or 3 months provided growth factors are constantly replenished and the cell concentration remains high ($>10^6$/ml). However, for unknown reasons after some weeks recovery of clonogenic cells falls off, exponential growth ceases, and cells die. The cells of patients with rapidly progressive disease generally proliferate *in vitro* at the highest rates for the longest times.

3.3 **Established cell lines**

Few established leukaemic lines are available. In a systematic attempt to develop lines from patients with leukaemia we grew cells at high cell concentrations in Iscove's medium, FCS, and PHA-LCM or tumour cell line CM (GCT, MoT 5637) using culture techniques given in *Table 10* (18). Eight lines were established. Five of them grew exponentially spontaneously in medium and FCS; three initially required a source of CSA and one remains GM-CSF dependent. Three ANLL and five ALL lines emerged under these conditions. Establishment of lines and prolonged growth *in vitro* tended to occur in patients with unfavourable karyotypes and with disease poorly responsive to treatment. However, the factors that allow some cells to grow exponentially indefinitely remain a mystery.

Table 10. Suspension culture of cells in acute leukaemia.

Materials

T-depleted leukaemic cells (*Table 4*)
Iscove's medium
Foetal calf serum
PHA-LCM or tumour cell line CM (*Table 8*)
25-ml Falcon flasks

Method

1. Place 5 ml of cells at 10^6/ml–5 × 10^6/ml in medium, 20% FCS with or without 10% conditioned medium.
2. Close flask tightly and place in 37°C incubator.
3. Remove aliquots for counting and for colony-formation assay (*Tables 5* and *6*).
4. Resuspend cells weekly to at least 10^6/ml on fresh medium and CM either by dilution or concentration. If cell number is $>2 × 10^7$/ml, it is necessary to subdivide more frequently.

3.4 **Virally transformed leukaemic cell lines**

Established lines do grow readily when transforming viruses such as EBV, HTLV-TI and HTLV-II are present in the malignant cells. Marrow and peripheral blood from normal healthy adults who have had infectious mononucleosis contain cells with the EBV genome and these cells give rise to lines in 40% of persons who have had infectious mononucleosis in the recent past or in a small percentage who had had it in the distant past (13). Because these spontaneous lines can be confused with malignant lines derived from the leukaemia; it is necessary to ascertain the leukaemic origin of pre-B lines by means such as karyotype and demonstration of the absence of the EBV genome. On the other hand, EBV could be viewed as a means to promote growth of most pre-B ALL *in vitro*. However, most pre-B ALL cells do not have EBV receptors and cannot be infected with a transforming strain of EBV such as B-95-8 virus; thus, EBV is usually not a confounding factor.

HTLV-I allows the outgrowth of malignant T-cell lines from some patients with leukaemia of mature T-cell phenotype by inducing the IL2 receptor. These lines grow in RPMI media and 10% FCS. The virus from HTLV-II infected line Mo-T can also transform normal T-lymphocytes (11, 12). Care must be taken to avoid outgrowths of normal T-lymphocytes when Mo-T medium is used. Boiling the medium for 5 min inactivates the virus but not the GM-CSF.

4. LIMITATIONS OF SYSTEMS FOR GROWING HUMAN LEUKAEMIA CELLS *IN VITRO*

Techniques for growing human leukaemic cells *in vitro* are imperfect: they do not predict which cells, for colonies *in vitro*, multiply in suspension, and which will form established lines. There is tremendous interpatient variability in growth potential *in vitro*. Growth parameters do not correlate with the age or sex of the patient, with lineage, morphologic and immunologic phenotype, or karyotype of the cells. In most, but certainly not all studies, the ability to grow in suspension roughly correlates with rapid clinical progression and poor prognosis, and as does the ability to form secondary and tertiary colonies *in vitro* (42). Perhaps this correlation indicates that these cells have intrinsic information that enhances growth *in vitro* as well as *in vivo*.

Not only is there interpatient variability but there is also tremendous subpopulation heterogeneity within the leukaemic population of a given individual. Leukaemia is a clonal disease arising from a single cell. However, the disease we recognize is comprised of billions of progeny of this cell. For the most part, the progeny have limited proliferative potential. Monoclonal antibody studies show that the population that forms colonies *in vitro* has a different phenotype from that of the majority of cells (43). Gradient density separation of leukaemic cells indicates that the clonogenic cells have a lighter density than the majority of cells. Furthermore, the clonogenic cells may have different growth requirements than the whole leukaemic population and there may be subpopulations of clonogenic cells with specific growth requirements (44, 45). These requirements are complex and differ from those of normal cells. For example, feeder layers and PHA-LCM are necessary for growth of many leukaemic cells and they cannot be replaced entirely by the single recombinant or purified haematopoietic growth factors currently available. Nonetheless, the available culture systems are sufficient to test all the components of feeder layers and PHA-LCM as they are identified and cloned. Ultimately it will be possible to define the conditions for growth of each patient's leukaemia and each leukaemic subpopulation.

5. REFERENCES

1. Moore, M. A. S., Spitzer, G., Williams, N., Metcalf, D. and Buckley, J. (1974) *Blood,* **44**, 1.
2. Coulombel, L., Kalousek, D. K. and Eaves, C. J. *et al.* (1983) *N. Engl. J. Med.,* **308**(25), 1493.
3. Chang, J., Morgenstern, G. M. and Deakin, D. *et al.* (1986) *Lancet,* **I**, 294.
4. Metcalf, D. (1984) *Clonal Culture of Hemopoetic Cells: Techniques and Applications.* Elsevier, New York.
5. Pike, B. L. and Robinson, W. A. (1970) *J. Cell. Physiol.,* **76**, 77.
6. Griffin, J. D. and Löwenberg, B. (1986), *Blood,* **68**(6), 1185.
7. Till, J. E. and McCulloch, E. A. (1980) *Biochim. Biophys. Acta,* **605**, 431.
8. Nara, N. and McCulloch, E. A. (1985) *Blood,* **65**, 1484.
9. Ferrero, D. and Rovera, G. (1984) *Clin. Haematol.,* **13**, 461.
10. Poiecz, B. J., Ruscetti, F. W., Mier, J. W., Woods, A. M. and Gallo, R. (1980) *Proc. Natl. Acad. Sci. USA,* **77**, 6815.
11. Chen, I. S. Y., Quan, S. G. and Golde, D. W. (1983) *Proc. Natl. Acad. Sci. USA,* **80**, 7006.
12. Miller, G. (1975) *J. Infect. Dis.,* **130**, 187.
13. White, L., Meyer, R. R. and Benedict, W. F. (1984) *J. Natl. Cancer Inst.,* **72**(5), 1029.
14. Boyum, A. (1977) *Lymphology,* **10**, 71.
15. Swart, K., Hagemeijer, A. and Löwenberg, B. (1982) *Blood,* **59**(4), 816.
16. Minden, M. D., Buick, R. N. and McCulloch, E. A. (1979) *Blood,* **54**, 186.
17. Young, D. C. and Griffin, J. D. (1986) *Blood,* **68**(5), 1178.
18. Lange, B., Valtieri, M. and Santoli, D. *et al.* (1987) *Blood,* **70**(1), 192.
19. Vellenga, E., Young, D. C. and Wagner, K (1987) *Blood,* **69**, 1771.
20. Miyauchi, J., Kelleher, C. A. and Yang, Y. C. *et al.* (1987) *Blood,* **70**, 657.
21. Mitjavila, M. T., Villeval, J. L. and Cramer, V. P. *et al.* (1987) *Blood,* **70**, 965.
22. Swart, K. and Löwenberg, B. (1984) *Cancer Res.,* **44**, 657.
23. Löwenberg, B., Swart, K. and Hagemeijer, A. (1980) *Leuk. Res.,* **4**, 143.
24. Dicke, K. A., Spitzer, G. and Ahearn, M. J. (1976) *Nature,* **259**, 129.
25. DiPersio, J. F., Brennan, J. F. and Lictman, M. A. *et al.* (1980) *Blood,* **58**, 717.
26. Fogh, J. (1978) *Natl. Cancer Inst. Monogr.,* **49**, 5.
27. Gasson, J. C., Weisbart, R. H. and Kaufman, S. E. *et al.* (1984) *Science,* **226**, 1339.
28. Welte, K., Platzer, E. and Lu, L. *et al.* (1985) *Proc. Natl. Acad. Sci. USA,* **82**, 1526.
29. Delwel, R., Dorssers, L., Touw, I., Wagemaker, G. and Löwenberg, B. (1987) *Blood,* **70**, 333.
30. Griffin, J. D., Young, D. and Herrmann, F. *et al.* (1986) *Blood,* **67**, 1448.
31. Kelleher, C., Miyauchi, J. and Wong, G. *et al.* (1987) *Blood,* **69**, 1498.
32. Metcalf, D. (1986) *Blood,* **67**, 257.
33. Salmon, S. E. and Buick, R. N. (1979) *Cancer Res.,* **39**, 1133.
34. Dube, I. D., Eaves, C. J., Kalousek, D. K. and Eaves, A. C. (1981) *Cancer Genet. Cytogenet.,* **4**, 157.
35. McCulloch, E. A., Buick, R. N. and Curtis, J. E. *et al.* (1981) *Blood,* **53**, 105.
36. Izaguirre, C. A., Curtis, J., Messner, H. and McCulloch, E. A. (1981), *Blood,* **57**, 823.
37. Touw, I., Delwel, R., Bolhuis, R., van Zanen, G. and Löwenberg, B. (1985) *Blood,* **66**, 556.
38. Uckun, F. M., Fauci, A. S. and Heerema, N. A. *et al.* (1987) *Blood,* **70**, 1020.
39. Touw, I., Delwel, R., van Zanen, G. and Löwenberg, B. (1986) *Blood,* **68**, 1088.
40. Elias, L. and Greenberg, P. (1977) *Blood,* **50**, 263.
41. Miyauchi, J., Kelleher, C. A. and Yang, Y. C. *et al.* (1987) *Blood,* **70**, 657.
42. McCulloch, E. A., Curtis, J. E., Messner, H. A., Senn, J. S. and Germanson, T. P. (1982) *Blood,* **59**, 601.
43. Lange, B., Ferrero, D. and Pessano, S. *et al.* (1984) *Blood,* **64**, 693.
44. Löwenberg, B., Hagenmeijer, A. and Swart, K. (1982) *Blood,* **59**, 641.
45. Wouters, R. and Löwenberg, B. (1984) *Blood,* **63**, 684.

CHAPTER 5

Methods for clonal growth
and serial cultivation of normal human
epidermal keratinocytes and mesothelial cells

JAMES G. RHEINWALD

1. INTRODUCTION

All epithelial tissues of vertebrate animals have the following structural properties in common: they are formed of closely packed cells which are in direct contact with one another without intervening extracellular matrix materials; they have a free surface which faces either the external environment or an internal compartment that is not continuous with the circulation; and they contain keratin-type inter-mediate filaments as a major component of the cytoskeleton. Apart from these common features, the epithelia are otherwise very diverse. They include glandular tissues such as the pituitary, liver, and mammary; transporting tissues such as the kidney; ciliated, mucous-secreting tissues such as the tracheobronchus; and protective barrier tissues such as the bladder urothelium, cornea, and epidermis.

The cultivation of epithelial cells *in vitro* is essential to research directed at understanding the mechanisms by which the growth of these various tissues is regulated and their differentiated functions performed. The ability to expand pure populations of epithelial cells in culture makes possible the convenient isolation of large amounts of cell type-specific proteins and mRNAs, enabling the molecular characterization of cell and tissue differentiation. Clonal growth and serial subcultivability of cells permits somatic cell genetic experimentation and precise characterization of cell type-specific responses to mitogens, growth inhibitors and differentiation inducers. Furthermore, the ability to culture both normal and malignant cells of the same tissue type is essential for identifying changes in cell regulation that occur in malignant transformation.

My own interest in epithelial cell culture began in the early 1970s when, as a graduate student in cell biology, I found it interesting that human fibroblasts grew so well in culture while only short-term, high density cultures had been achieved for any epithelial cell type, including those such as epidermal and intestinal epithelial cells which divide every several days or weeks *in vivo*. In 1975 Howard Green and I reported the discovery that epidermal keratinocytes require special growth factors which are not needed by fibroblasts and are not present in serum, and that keratinocytes in culture continue to be subject to a terminal differentia-tion programme which is continuously removing cells permanently from the cell cycle (1, 2). This differentiation programme must be countered by maintaining the cells in the presence of factors and under conditions which minimize the rate at which proliferating cells become committed to terminal differentiation (3, 4). The original culture conditions have since been improved upon by the identification of additional factors that function as mitogens or differentiation inhibitors for this cell type (5–7). The current version of the culture system is described in detail in Section 4 of this chapter.

The epidermal cell culture system described here is also excellent for promoting the growth of keratinocytes (stratified squamous epithelial cells) from the oral cavity (8; K. Lindberg and J. G. Rheinwald, in preparation), oesophagus (9, 6), exocervix (10, 6) and cornea (9, 11, 12), and for tracheobronchial epithelial cells (13), conjunctival epithelial cells (11), bladder urothelial cells (6, 14) and mammary epithelial cells (15, 6). Not all epithelial cell types are serially propagable in culture in this system, however, or, in fact, in any system published to date. These include cells such as intestinal epithelial cells and hepatocytes, for which only short-term culture has been achieved. Certain other epithelial cell types grow well under conditions very different from and simpler than those necessary for the growth of keratinocytes and the other cell types listed above. An important example is the mesothelial cell, which my lab has studied extensively and for which we have identified an optimal growth medium (6, 16, 17). Methods for culturing human mesothelial cells are described in Section 5.

2. BASIC STRATEGIES AND CONSIDERATIONS

As mentioned above, a main consideration for maximizing the serial cultivability of epidermal cells and other terminally differentiating epithelial cells is that the cells must be kept in conditions that minimize the rate at which they commit irreversibly to terminal differentiation. For keratinocytes this requires subculturing at each passage before colonies become so large that cells in colony centres become overcrowded and inhibited from centrifugal migration (3, 18). Another requirement of successful epidermal cell culture is that the cell population is purified of contaminating dermal tissue fibroblasts which are usually present in cell suspensions derived even from carefully prepared skin samples. Because fibroblasts have much higher colony-forming abilities than epidermal cells and are not subject to terminal differentiation, small numbers of fibroblasts in the primary culture can overtake the epidermal cell population within several passages unless they are removed at an early stage. Methods are described in Section 4.4 for complete, selective removal of fibroblasts during the primary and secondary cultures.

Mesothelial cells are not subject to terminal differentiation, as far as it is known. This cell type therefore tends to exhibit a similar replicative lifespan and colony-forming efficiency at subculture whether grown in optimal or growth factor-deprived conditions, and whether or not they are cultured to high density before passage. The convenient source of normal human mesothelial cells described in Section 5.2—namely, ascites fluid or pleural effusion fluid (6, and literature cited therein)—does not contain contaminating fibroblasts. Thus a special procedure for purifying the mesothelial population of fibroblasts is not necessary. Both epidermal cells and mesothelial cells have distinctive morphologies, which makes their microscopic identification and distinction from other possible cell types rather certain. In addition, these cell types express high levels of certain intermediate filament proteins which are readily identifiable with indirect immunofluorescence using commercially available antibodies, thus ensuring the identification of these cell types in culture, as described in Sections 4.7 and 5.5.

3. CULTURE MEDIA, SUPPLEMENTS AND SOLUTIONS

Foetal calf serum (FCS) is obtained either from Hyclone, Inc. (Ogden, UT) or Sigma (St Louis, MO). Serum is purchased after a new lot has tested satisfactorily against a current lot known to promote high colony-forming efficiency, maximum growth rate, and maximum subcultivability of the desired cell type. Serum lots optimal for epidermal cells are not usually the same as those for mesothelial cells. (The two companies mentioned above have by far the highest proportion of suitable serum lots of all commercial suppliers of serum we have tried over the past 10 years.) Serum is purchased in lots large enough to last for 1–2 years, and is stored at $-20°C$.

Commercially prepared nutrient media formulations of DME, F12, M199, and MCDB105 (Sigma) are purchased in powdered form in lots large enough to last for 6–12 months and packets are stored at 4°C until use. Samples of nutrient medium lots are tested before purchase against current lots for growth promotion in the same way as serum is tested. All media are supplemented with 0.1 g/litre streptomycin and 0.1 g/litre penicillin (ICN Biochemicals). Serum is added to medium at the time it is used for feeding cells and the serum-supplemented medium is then stored in the refrigerator for no longer that 1 month.

Phosphate-buffered saline (PBS) is prepared by dissolving 80 g of NaCl, 2 g of KCl, 11.5 g of Na_2HPO_4 (or 21.4 g of $Na_2HPO_4 \cdot 7H_2O$) and 2 g of KH_2PO_4 in 10 litres of water and adjusting the pH to 7.35 with 1 M HCl or 1 M NaOH. (The pH of trypsin and EDTA solutions in PBS is readjusted to pH 7.35 after preparation.)

Ca^{2+}-free Hepes-buffered Earle's salts (HBES) is prepared by dissolving 64 g of NaCl, 4 g of KCl, 37 g of $NaHCO_3$, 125 g of $NaH_2PO_4 \cdot H_2O$, 2 g of $MgSO_4 \cdot 7H_2O$, 0.001 g of $Fe(NO)_3 \cdot 9H_2O$, 60 g of Hepes (Sigma) and 10 ml of 0.5% Phenol Red solution (Flow Labs) in 10 litres of water. The HBES solution is stored frozen and is thawed and filter-sterilized as needed.

Mitomycin C (Sigma) is prepared fresh each week by dissolving a 2 mg vial in 10 ml of HBES. This ($50\times$ for 4 μg/ml) solution is filter-sterilized and stored refrigerated. It is kept shielded from light when not in use.

3.1 Growth factors and hormones

Hydrocortisone (Behring Corp.) is first dissolved in 100% ethanol to a concentration of 5 mg/ml. This solution can be stored for a year at $-20°C$. It is dissolved 1/125 in HBES + 5% FCS to make a 40 μg/ml stock solution ($100\times$ for 0.4 μg/ml) or 2/25 to make a 400 μg/ml (1000× for 0.4 μg/ml) solution. This is filter-sterilized and stored at −20°C for as long as 1 year.

Epidermal growth factor (EGF) is dissolved at 1 μg/ml or 10 μg/ml in HBES + 5% FCS and filter-sterilized to make 100× or 1000× for 10 ng/ml stock solutions, which are stored at −20°C for as long as 1 year.

Cholera enterotoxin (Schwarz–Mann) is dissolved in sterile water at 1 mg/ml (10^{-5} M) and is stored at −20°C. This concentrated solution is diluted 1/1000 in HBES + 20% FCS and filter-sterilized, making a 100× for 10^{-10} M stock solution, which is stored at −20°C for as long as 6 months.

Insulin (Bovine pancreas, Sigma) is dissolved in 0.005 M HCl at 5 mg/ml and is filter-sterilized through a low protein-binding filter. This (1000× for 5 μg/ml solution) is stored at 4°C for as long as 2 weeks.

Adenine is dissolved at 243 mg/100 ml or 24.3 mg/10 ml in 0.05 M HCl with stirring for 1 h. This (100× for 1.8×10^{-4} M) stock solution is filter-sterilized and stored at −20°C for as long as 1 year.

Trypsin (Hog pancreas 1:300, ICN Biochemicals) is prepared by dissolving 1 g trypsin + 1 g glucose + 3 ml 0.5% Phenol Red solution (Flow Labs) in 1 litre of PBS (pH 7.35). The solution is filter-sterilized, divided into 20-ml aliquots, and stored at −20°C for as long as 6 months.

EDTA (ICN Biochemicals) is dissolved in PBS at 0.2 g/litre (pH 7.35), filter-sterilized, and stored at 4°C for as long as 1 year.

4. EPIDERMAL KERATINOCYTE CULTURE

4.1 Culture medium

Keratinocyte cultures are fed with a medium consisting of Dulbecco's modified Eagle's medium (DMEM)/F12 (3:1 v/v) supplemented with 5% FCS, 0.4 μg/ml hydrocortisone (HC), 10 ng/ml EGF, 10^{-10} M cholera toxin, 1.8×10^{-4} M adenine, and 5 μg/ml insulin (7). Some researchers supplement this medium with 5 μg/ml transferrin and 5 μg/ml triiodothyronine, but we have never detected a stimulatory effect of these agents in this culture system. Some researchers also use a larger serum supplement of 10 or 20%, but we have detected no enhancement of growth above 5% FCS. In fact, in the medium described above, nearly maximal growth rate and colony-forming efficiency is obtained with a serum supplement as low as 0.5% and considerable growth occurs with 0.2% FCS (our unpublished observations), indicating that keratinocytes are not dependent upon any serum mitogen that the 'cocktail' of factors listed above does not provide or replace.

4.2 The 3T3 fibroblast feeder layer

We have always used the random-bred Swiss mouse 3T3 line developed by Todaro and Green (19), which is different from the NIH Swiss 3T3 line widely used for DNA transfection studies. The NIH Swiss 3T3 and BALB/c-3T3 cell lines also work adequately as feeder cells. However, some versions of these lines which we have obtained do not support growth quite as well as a particular clone, 3T3J2, of the Todaro and Green line I isolated more than 10 years ago. Another clone, 3T3M1, a 6-thioguanine-resistant variant isolated in my laboratory for use in supporting keratinocyte growth during selection of induced *hprt−*mutants (7), is also an adequate feeder cell, but does not promote growth under normal conditions quite as well as 3T3J2.

The 3T3 cells are grown in DMEM + 10% calf serum (CS) and are maintained as continually dividing stocks for a span of 4–8 weeks before a new ampoule of cryopreserved cells is thawed to replace the old cultures. In our hands 3T3 cells eventually begin to grow more slowly, exhibit morphological heterogeneity, become unable to grow to a fully confluent monolayer and function less well in supporting keratinocyte growth. 3T3 stocks are subcultured weekly as follows.

On Thursday, a recently confluent 100 mm culture that had been plated the previous week is disaggregated to a single cell suspension by incubating it with 0.1% trypsin for 10 min at 37°C. The cells are resuspended in 8 ml of DMEM + 10% CS and 100 mm dishes are plated with 1, 0.5, 0.25, and 0.12 ml such that cultures are plated at 1/8, 1/16, 1/32, and 1/64 confluent density. The highest density platings will become fully confluent on Monday and others later in the week. Dense cultures are refed or are treated on Monday, and the other cultures are refed the day before they become confluent and are to be prepared as feeders.

To prepare 3T3 cells as feeders, confluent cultures (or trypsinized cells resuspended in medium) are placed within a ^{60}Co source and irradiated with 6000 rads. Alternatively, Mitomycin C is added to the medium of confluent 3T3 cultures at 4 μg/ml for 2 h at 37°C. The cultures are then rinsed twice with trypsin, the cells disaggregated with trypsin, and replated at about one-third their saturation density ($\sim 10^6$ per 100-mm dish or 3.5×10^5 per 60-mm dish) in DMEM + 10% CS.

Three hours to 2 days later, the medium is replaced with epidermal culture medium and epidermal cells are added. (If desired, mitomycin-treated or irradiated 3T3 cultures can be fed fresh DMEM + 10% CS medium and returned to the incubator for several days before being trypsinized and replated for use as feeders.)

4.3 Initiating cultures from skin

Samples of surgically excised, full-thickness skin from circumcisions or from cosmetic plastic surgery procedures, such as reduction mammoplasties, are placed aseptically into sterile culture medium and transported to the laboratory, where they are kept at room temperature or at 4°C for as long as 24 h before processing. The rate of loss of viability with time after surgical removal has not been determined for epidermal cells, so disaggregation is initiated as soon as possible. Skin pieces of 1–4 cm^2 epidermal surface area are inverted in serum-free medium in a 100-mm culture dish. As much of the dermis as possible is removed using a forceps and curved iris scissors, and is either discarded or is used to generate fibroblast cultures. The epidermis and adherent uppermost portion of the dermis is transferred to another dish without medium and is reduced to fragments of less than 1 mm^3 by repeated mincing with a curved iris scissors. The minced material is then transferred to a 25–50-ml suspension culture vessel equipped with a suspended, magnetic stir-bar which has several millimetre clearance from the bottom of the vessel. 10–15 ml of a 1 : 4 mixture of 1% collagenase (Worthington CLS II) and 0.25% trypsin in PBS is added and the vessel is then placed in a 37°C room on a magnetic stirring platform set to 60–120 r.p.m. At 40 min intervals the vessel is brought back to the culture hood and fragments are allowed to settle. The supernatant containing single cells and small clusters is drawn off and added to a test tube containing an equal volume of DMEM + 10% CS. An aliquot (10–15 ml) of trypsin/collagenase solution is added back to the pieces remaining in the disaggregation vessel, which is then returned to the 37°C room for another 40-min digestion period. Suspended cells are harvested and fresh trypsin/collagenase solution is added back to the remaining pieces every 40 min for a total of 2–3 h.

The cells recovered at each harvest are resuspended in complete epidermal cell medium, counted, and plated at densities of 10^5 and 3×10^5 cells per 100-mm dish with 3T3 feeder cells. 50 μg/ml gentamicin (Gibco) and 50 U/ml mycostatin (Squibb) are also added to the medium for the first week of culture to reduce the probability of bacterial, yeast or fungal contamination. Epidermal colony forming efficiencies from released cells usually range from 0.1–2% in the primary culture. Cultures are refed 4 days after plating and every 3 days thereafter until subculture.

4.4 Purifying keratinocyte cultures of contaminating fibroblasts

A near-confluent 3T3 feeder layer is very effective at inhibiting human fibroblast growth, but fibroblasts can divide in any area of the dish surface that becomes denuded of feeder cells. Six or 7 days after plating, when epidermal colonies in the primary culture have reached an average size of 50–200 cells but before neighbouring colonies have merged, the medium is removed and the cell layer is rinsed with a solution of 0.02% EDTA in PBS. The cultures are then incubated with EDTA at room temperature in the hood for 30 sec, and then the cell layer is sprayed repeatedly and rather vigorously with pipette streams of the EDTA solution for a period of 30 sec as the dish is rotated to ensure that the entire area of the dish is sprayed. This procedure dislodges the 3T3 feeder cells and any human dermal fibroblasts, but leaves epidermal colonies firmly attached.

The EDTA solution is then removed, 5 ml of fresh EDTA solution is added, and the culture is quickly inspected under an inverted microscope to assess the completeness of fibroblast removal. Any remaining areas of fibroblast 3T3 feeder cells are dislodged by further pipetting. The dish is then rinsed with four 5-ml changes of serum-free medium and complete epidermal medium is added back. Fresh irradiated or mitomycin-treated 3T3 feeder cells are then added and the culture is returned to the incubator to grow further until it is ready for subculture. The selective EDTA removal of 3T3 feeder fibroblasts and any living human dermal fibroblasts can be repeated 5–7 days after plating the secondary culture if any human fibroblasts are still evident. However, EDTA removal of the 3T3 feeder cells is also the first step of the epidermal cell subculture procedure (described in Section 4.6), so if the cells are subcultured before neighbouring colonies have merged, cell populations that originally contained even a large number of dermal fibroblasts are usually completely free of them by the secondary culture.

4.5 Growth of colonies and retention of the clonogenic cell fraction

Human epidermal keratinocytes form tightly adherent epithelioid colonies which, by the time they have attained a size of 50–200 cells, have begun to stratify. The stratified cells are very flat and difficult to see, but they cover all but the peripheral three to six cell diameters of the colony. Possibly because the central basal cells, which are covered up, become starved of nutrients or mitogens, or because peripheral basal cells in large colonies become unable to migrate outward rapidly enough to accommodate a doubling in cell number and colony area each day (18), growth rate begins to slow by 8–11 days, even in cultures that are subconfluent (3, 18). Epidermal cells then either become committed at a higher rate to terminal differentiation or else they become less able to withstand subculture, because they display a substantial decrease in colony-forming ability (i.e. the percentage of plated cells yielding large, progressively growing colonies in the next passage). This relative decrease in subcultivability occurs at each passage if the cells are allowed to grow for more than 8–10 days after plating before they are disaggregated and subcultured. Thus if maximum cell mass for biochemical analysis is not the goal, but rather maximum yield of proliferative cells for subculture or cryopreservation, it is clear that from a 100-mm culture a yield of 3×10^5 cells having a colony-forming potential of 8% is superior to a yield of 3×10^6 cells having a colony-forming potential of 0.5% from the same culture if it were allowed to grow for an additional 4 days before subculture. The colony-forming ability of cells at any passage decreases steadily after about 8 days after plating and drops precipitously after the colonies have merged to form a confluent cell layer.

Newborn foreskin epidermal cells usually exhibit a replicative lifespan of 80–100 cell generations over a 3–4 month period of continuous serial cultivation before they become senescent. (See ref. 20 for a brief review of the literature on senescence of normal human cells and calculation of replicative lifespan.) Epidermal cells from children or adults exhibit a shorter lifespan (40–70 cell generations) than newborn cells and lower colony-forming abilities throughout the lifespan.

Maximal colony-forming potential of epidermal cell strains throughout the lifespan is obtained by plating cells at 3×10^4, 10^5 and 3×10^5 per 100-mm dish (or 10^4, 3×10^4 and 10^5 cells per 60-mm dish) and choosing a healthy, preconfluent culture to subculture every 7–8 days. The colony-forming ability of a cell population can be checked by plating cells at 100, 300 and 10^3 cells per 100-mm dish and growing them for 14 days. The cultures are then fixed, stained, and examined to count the number of large colonies/number of cells plated. One can expect colony-forming abilities of 2–20% from newborn foreskin cells during the first four or five of these weekly passages and 0.5–7% for adult cells. These percentages usually decrease to 0.1–1% as cells near senescence.

A good strategy for maintaining a particular cell strain for nearly indefinite use is to cryopreserve at least a dozen ampoules of cells each from the primary culture, from the secondary culture, and from the third passage culture. Experiments requiring cells for biochemical analysis can use cultures initiated from third passage freezings, while most experiments requiring cells with long-term growth potential and high colony-forming abilities can use cultures initiated from second passage freezings. As third passage ampoules become depleted, a second passage ampoule can be thawed, expanded in culture, and refrozen to replenish the third passage stock; the same can be done with primary ampoules to replenish the second passage stock. Cryopreservation methods are described in Section 6.2.

4.6 Subcultivation

Cultures are rinsed with 0.02% EDTA in PBS and the solution is pipetted vigorously to dislodge the 3T3 feeder cells, as described in Section 4.4 above. Dislodged cells are aspirated and discarded and the epidermal cells are then covered with a 1:1 mixture of 0.02% EDTA and 0.1% trypsin solutions, using total volumes of 2.5 ml per 60-ml dish or 6 ml per 100-mm dish. The culture is placed at 37°C for 20 min and then is examined under the inverted microscope. The cells should be rounded up, detached, and floating as single cells or in small, loosely adherent aggregates. If not, the culture is returned to 37°C for another 10 min. The detached cells are gently pipetted up and down several times and re-examined under the microscope to ensure that the majority of the population has been reduced to a single cell suspension. The cell suspension is then diluted with an equal volume of DMEM + 10% CS, pelleted by low speed ($\leqslant$500 r.p.m.) centrifugation for 5 min in a tabletop centrifuge and resuspended in complete epidermal cell growth medium for further use.

4.7 Identifying keratinocytes in culture

Keratinocytes have a distinctive colony morphology in this culture system. Cells form tightly adherent, epithelioid colonies with one or two flat, stratified cell layers discernible with phase-contrast microscopy by the faint borders of the stratified cells toward the colony centres. Small colonies are often hard to see against the background of 3T3 feeder cells, but are easily identified after selective removal of the feeder cells with EDTA, as described in Section 4.4 above.

Cultured epidermal cells produce abundant amounts of keratins 5, 6, 14, 16 and 17, which are recognized via indirect immunofluorescence by monoclonal antibodies and conventional antisera available commercially, or which can be identified electrophoretically (mol. wt 46–58 kd in SDS–polyacrylamide gels), as reviewed in ref. 21. Stratified epidermal cells in colony centres also contain the terminal differentiation-associated proteins involucrin (22, 23) and epidermal transglutaminase (24). Specific antibodies can be used to detect the keratinocyte-specific proteins by indirect immunofluorescence.

4.8 An alternative culture system for human epidermal cells

Normal human epidermal cells also grow well from low density platings through several serial passages in a different medium, originally developed by Peehl and Ham at the University of Colorado (25). A slight modification of this medium (26, 27), MCDB153, is available commercially (Clonetics Corp.). This system uses a nutritionally optimized medium formulation supplemented with 70 μg/ml crude bovine pituitary extract to replace the functions of the 3T3 fibroblast feeder layer.

In our hands, most of the lots we have obtained of the commercial preparation of this medium stimulated good short-to-medium term (one or two passages) growth although with lower colony-forming efficiencies than obtained in the 3T3 feeder layer system.

5. MESOTHELIAL CELL CULTURE

The mesothelial cell is the simple squamous epithelial cell type that covers all surfaces exposed to the inner body cavities, that is the pleura, pericardium, and peritoneum. Mesothelial cells are derived developmentally from the embryonic mesoderm. About 8 years ago, while we were trying to culture metastatic ovarian carcinoma cells from ascites fluid using the 3T3 feeder layer system, we found that the most prolific cells from these samples were normal mesothelial cells (6). We found that their optimal growth medium was quite different from that of epidermal cells (16, 17). We have submitted two normal human peritoneal mesothelial cell strains, LP-9 and LP-3, to the Cell Repository at the Coriell Institute for Medical Research, Camden, NJ, for general distribution.

5.1 Culture medium for human mesothelial cells

The optimal medium for normal human mesothelial cells (16, 17, 28) is a 1:1 mixture (v/v) of M199 and MCDB105 medium (Sigma) supplemented with 5–10 ng/ml EGF, 0.4 μg/ml HC, and 5–20% FCS (note: we find that 7% serum yields nearly maximal growth rates). Cells grow satisfactorily in M199 medium, but grow very slowly in F12/DMEM medium. High density cultures can be maintained well in F12/DMEM medium, however.

5.2 Initiating cultures from peritoneal or pleural fluids

Ascites fluid or pleural effusion fluid (tapped aseptically from patients who have accumulated this fluid pathologically as a consequence of metastatic cancer in one of these body cavities) is collected in a sterile bottle containing heparin by clinical staff and is processed in the lab within an hour of collection. (Fluids from mesothelioma patients are avoided so as not to risk culturing abnormal mesothelial cells.) The closed bottle is inverted several times to ensure that cells are suspended, and 45-ml aliquots are transferred to four 50-ml disposable centrifuge tubes. Cells are pelleted by 10 min of low speed (500 r.p.m.) centrifugation in a tabletop centrifuge. The pellets, containing erythrocytes, macrophages, cancer cells, and mesothelial cells, are resuspended in DME + 10% CS and pooled in a final volume of 10 ml. Nine ml are set aside for cryopreservation. The remaining cells are plated at 0.1, 0.3, and 0.6 ml per 100-mm dish in complete mesothelial cell medium.

One day after plating, the dishes are rinsed gently but thoroughly with several changes of serum-free medium to remove erythrocytes and other non-attached debris, and the cultures are refed with complete medium. After 4–7 days after plating, dispersed colonies of rapidly proliferating cells with stubby fibroblastoid morphology become evident. Colony-forming mesothelial cells are usually present in ascites and pleural effusion fluids at concentrations ranging from about 20 to 10^3 cells per ml; thus the cell suspensions concentrated in culture medium for cryopreservation or plating in primary culture typically contain 100–5000 colony formers per ml. Of 20 ascites and pleural effusion samples we have cultured, which contained metastatic ovarian, breast, colon, and lung carcinoma cells as well as mesothelial cells, only one sample yielded rapidly and progressively growing cancer cells. This sample could not be used to obtain pure mesothelial cell populations. In all other cases the cancer cells did not grow, the macrophages degenerated, and essentially pure mesothelial populations were present at the end of the primary culture 8–14 days after plating.

5.3 Subcultivation

Mesothelial cells grow to high saturation density ($>10^7$ cells per 100-mm dish) in complete medium, and they degenerate because of medium depletion if they are not fed every day once they have reached this density. Although mesothelial cells do not terminally differentiate or otherwise convert to a non-subcultivable state with increasing cell density, healthier subcultures with higher plating efficiency are obtained if cells are passaged when they are preconfluent or have just reached confluence.

Cultures are disaggregated to single cells by a 10-min incubation with 0.1% trypsin in PBS at 37°C. The trypsinized cell suspension is mixed with an equal volume of medium plus serum, centrifuged, and resuspended in complete medium for counting and subculture.

Normal adult human mesothelial cells exhibit a replicative lifespan of 40–50 cell generations and a colony-forming ability of 5–25% during the first two-thirds of their lifespan. Stock cultures are easily maintained for use throughout the week by plating cells at 3×10^3, 10^4, 3×10^4, 10^5, 3×10^5 and 10^6 per 100-mm dish in complete medium. Cultures do not require refeeding until they near confluence or 1 week has elapsed since the last feeding. Every 7–10 days one of these dishes is subcultured when it nears confluence and is plated again at densities ranging from 3×10^3 to 10^6 per dish. The cell population begins to accumulate large, flat cells and growth begins to slow about one passage before complete senescence.

Primary and secondary cultures are cryopreserved so that the frozen ampoules of the original, uncultured ascites or plueral fluid cells are only thawed for the purpose of restoring depleted reserves of frozen primary and secondary passage cells.

5.4 Preparing 'differentiated' mesothelial cell populations

Mesothelial cells normally exist in a non-dividing state *in vivo*, where they have a high keratin content. When they are placed in culture in complete medium, they assume a quasi-fibroblastoid morphology, greatly reduce their rate of keratin synthesis, and become depleted of keratin filaments (16). They form locally oriented, parallel arrays of cells as they become confluent, and then overgrow to form a dense network several cells deep, similar to the behaviour of normal human dermal fibroblasts. This behaviour is growth-rate-dependent, and is particularly EGF-dependent.

If a normal human mesothelial cell culture is deprived of EGF when it is one-quarter or less confluent, the cells slow their growth rate substantially, become more flattened and stop dividing when they reach a single cell monolayer, which is reminiscent of their *in vivo* morphology. Under these conditions they also increase their rate of keratin synthesis and regain a high keratin content. EGF-deprived, confluent mesothelial cultures are not identical to *in vivo* mesothelium, however, as they retain a substantial vimentin content (16) and they secrete high levels of a plasminogen activator inhibitor PAI-1, which we had first identified as mesosecrin (29).

EGF-deprived cultures are prepared by plating 3×10^5 cells per 100-mm dish in complete medium lacking EGF. Alternatively, cells can be plated at 10^5 per dish in complete medium and 2 days later the cultures can be rinsed well and refed with medium lacking EGF. Cultures are then refed twice weekly until confluent. Cultures are then fed three times per week. They remain healthy in appearance and continue to synthesize large amounts of keratin for at least 2 months under these conditions.

5.5 Identifying mesothelial cells

Rapidly proliferating human mesothelial cells exhibit a distinctive morphology in culture (16). The grow in a dispersed fashion—not closely adherent, as many other epithelial cell types, because they do not tend to form desmosomes—but they are not as long and spindly as human fibroblasts. Mesothelial cells often form a broad ruffled membrane along one entire side of the cell body (their leading edge of migration). When the cells are deprived of EGF at subconfluent densities, the cells become flattened and form a characteristic monolayer of flattened cells (16).

In culture, mesothelial cells coexpress the intermediate filament proteins vimentin and the four simple epithelial keratins: K7(M_r 55 kd), K8(M_r 52 kd), K18(M_r 44 kd) and K19(M_r 40 kd) (6, 16, 14). These proteins can be identified easily using commercially available antibodies and indirect immunofluorescence microscopy. During rapid growth, many cells are very low in or devoid of keratins; they usually retain a low but detectable level of K8 and K18 but often become K7- and K19-negative. Cell populations are usually heterogenous in keratin content during rapid growth but are of uniformly high keratin content at saturation density in the absence of EGF. The only other normal human cell type that has been reported to reversibly modulate its keratin content this way is the Type II kidney epithelial cell (30, 14), so this behaviour is confirmatory for the identity of a cell cultured from pleural, peritoneal, or pericardial fluid. Human mesothelial cells in culture also secrete high levels of a M_r 46 kd plasminogen activator inhibitor, PAI-1/mesosecrin (29). This protein is also produced at equivalent levels by endothelial cells and kidney epithelial cells in culture, but at substantially lower levels by fibroblasts, and at almost undetectable levels by many other epithelial cells types (29). Thus its presence in abundance, as detected by electrophoretic examination of the culture medium of cells labelled with [^{35}S]methionine helps confirm the identity of a cultured cell as a mesothelial cell.

6. GENERAL METHODS

6.1 Staining cultures

Cultures are rinsed very briefly with slowly running, cold tap water, and are fixed by the addition of 10% formalin in PBS (Fisher Scientific) for 10 min to as long as a week at room temperature. (Note that formaldehyde vapours are a health hazard, so pouring of the formalin solution and storage of fixed cultures should be done in a well-ventilated room or in a vented fume hood. Lids should be replaced on culture dishes after the addition of formalin to minimize vaporization.)

A 0.2% methylene blue solution is prepared by dissolving 1 g of the powdered dye in 500 ml of distilled water for 2 h at room temperature. The solution is then filtered through Whatman no. 1 filter paper and stored in a closed bottle. Formalin-fixed cultures are rinsed with two changes of slowly running tap water, and the methylene blue solution is then added for at least 10 min or for as long as a week. The used dye solution is poured back into the bottle and the stained culture dishes are rinsed with five changes of slowly running, cold tap water. Dishes are then turned faced downward, tapped firmly against absorbant paper to shake off excess water, and air-dried in an inverted position. The dye solution can be reused many times, but is discarded when it no longer stains cultures a deep blue colour.

6.2 Cryopreservation

Cultures chosen for cryopreservation should be slightly preconfluent and should be fed with fresh medium the day before freezing. Cultures are disaggregated to single cells in the usual way and resuspended in DMEM or F12/DMEM medium supplemented with 5–10% CS or FCS. Cells can be frozen at concentrations of 10^3 to 2×10^6 cells/ml. Either 10% glycerol or 10% dimethylsulphoxide (DMSO) is used as a cryprotectant.

Method 1. 20-ml aliquots of pure glycerol are sterilized by autoclaving in glass bottles and are stored refrigerated for as long as 1 year. Glycerol is warmed to 37°C and is pipetted directly into the cell suspension to a final concentration of 10% (e.g. 1 ml of glycerol is added to 9 ml of cell suspension in medium).

Method 2. A 20% DMSO solution in FCS (e.g. 10 ml of DMSO added to 40 ml of serum) is prepared by slowly adding pure DMSO (Fisher Scientific) with swirling to serum chilled to 4°C in an ice bucket. The DMSO/serum solution is refrigerated overnight, then filter-sterilized through a 0.2 μm filter and stored refrigerated for as long as 6 weeks. An equal volume of chilled DMSO/serum solution is added to the cell suspension, making the final DMSO concentration 10%.

Cell suspensions, either in glycerol or DMSO, are pipetted into screw-capped, plastic freezing vials (Nalgene) and are slowly frozen (-1°C/min) either by placing them in a styrofoam insulated chamber in the neck of a liquid nitrogen freezer (Union Carbide) for at least 3 h, or by placing them in a programmed freezing apparatus (Planar or Cryomed) set to cool the cells at a rate of $-1°C/min$ to a temperature of $-30°C$ and then at $-5°C/min$ to a temperature of $-80°C$. The frozen ampoules are then quickly transferred to racks or boxes and stored in liquid nitrogen vapours or immersed in liquid nitrogen. Preliminary comparisons in our lab indicate that cells recover equally well after freezing in glycerol and DMSO when they are cooled in a programmable apparatus, but when cooled in a styrofoam chamber, which does not provide a strictly controlled or reproducible cooling rate, they show a better percent survival with DMSO.

Ampoules are thawed by dropping them into 500 ml of 37°C water in a polyethylene beaker. (Note: a protective face shield is worn to protect against the rare event of ampoule explosion.) Immediately after thawing, ampoules are swabbed with 70% ethanol, the cap is removed and the cell suspension transferred with a Pasteur pipette to a sterile centrifuge tube. Nine millilitres of DMEM + 10% CS is added slowly with swirling at a rate of about one drop every 2 sec for the first 2 ml, then 0.5 ml every 2 sec for the next 3 ml, and a final addition of all the remaining medium at once. The cells are pelleted by centrifugation, resuspended in the appropriate culture medium and plated as desired.

6.3 Assessing proliferative potential

Diploid human cells cultured from normal tissue samples proliferate in culture for only a limited number of cell divisions before they succumb to an internal 'senescence' mechanism and permanently stop dividing (see ref. 20 and literature citations therein). For purposes of producing cell mass for biochemical studies, the important feature of this limited replicative lifespan is the total number of *population doublings* a cell strain undergoes from the time it is placed in culture until it becomes senescent. This is calculated simply as the sum of the doublings in cell number undergone by the cells at each passage:

$$\log_2 \frac{\text{number of cells at subculture}}{\text{number of cells plated}}$$

Because the proportion of proliferating cells in the inoculum plated in the primary culture is so low, growth that occurs in the primary culture is often ignored in this calculation. However, cell divisions are occurring and a significant part of the lifespan elapses in the primary culture and this growth differs among tissue isolates, so it is often impossible to make precise comparisons of replicative potential among cell strains using population doublings as the measurement.

Replicative potential is more precisely determined as *cell generations*, which is the total number of divisions undergone by the longest lived cells of the strain before they become senescent. This number can include the number of divisions that occurred in the primary culture, provided that the number of original colony-forming cells is known approximately. It also takes into consideration the fact that many cells that are plated at each passage either do not reinitiate growth or else give rise to only small, abortive colonies which do not contribute a significant proportion of the cells that have grown in the cultured population by the next time it is passaged. *Cell generations* is calculated as:

$$\log_2 \frac{\text{number of cells at subculture}}{(\text{number of cells plated}) \, (\text{colony-forming efficiency})}$$

7. ACKNOWLEDGEMENTS

It is a pleasure to acknowledge the contributions of my technicians, students and postdocs, particularly Therese O'Connell, Lynn Allen-Hoffmann, and Michael Beckett, to the development of the culture systems described here. I also wish to thank Howard Green, Yann Barrandon and Niyi Kehinde for many useful discussions about keratinocyte culture, and Kelly Havican for preparing the manuscript.

8. REFERENCES

1. Rheinwald, J. G. and Green, H. (1975) *Cell*, **6**, 317.
2. Rheinwald, J. G. and Green, H. (1975) *Cell*, **6**, 331.
3. Rheinwald, J. G. and Green, H. (1977) *Nature*, **265**, 421.
4. Rheinwald, J. G. (1979) *Int. Rev. Cytol.*, suppl., **10**, 25.
5. Green, H. (1978) *Cell*, **15**, 801.
6. Wu, Y. J., Parker, L. M., Binder, N., Beckett, M. A., Sinard, J. H., Griffiths, C. T. and Rheinwald, J. G. (1982) *Cell*, **31**, 693.
7. Allen-Hoffman, B. L. and Rheinwald, J. G. (1984) *Proc. Natl. Acad. Sci. USA*, **81**, 7802.
8. Taichman, L., Reilly, S. and Garant, P. R. (1979) *Arch. Oral Biol.*, **24**, 335.
9. Doran, T. I., Vidrich, A. and Sun, T.-T. (1980) *Cell*, **22**, 17.
10. Stanley, M. S. and Parkinson, E. K. (1979) *Int. J. Cancer*, **24**, 407.
11. Sun, T.-T. and Green, H. (1977) *Nature*, **269**, 489.
12. Schermer, A., Galvin, S. and Sun, T.-T. (1986) *J. Cell Biol.*, **103**, 49.
13. Lechner, J. F., Haugen, A., Autrup, H., McClendon, I. A., Trump, B. F. and Harris, C. C. (1981) *Cancer Res.*, **41**, 2294.
14. Rheinwald, J. G., O'Connell, T. M., Connell, N. D., Rybak, S. M., Allen-Hoffmann, B. L., LaRocca, P. J., Wu, Y.-J. and Rehwoldt, S. M. (1984) In *Cancer Cells 1/The Transformed Phenotype*. Cold Spring Harbor Laboratory Press, New York, p. 217.
15. Taylor-Papadimitriou, J., Shearer, M. and Stoker, M. G. P. (1977) *Int. J. Cancer*, **20**, 903.
16. Connell, N. D. and Rheinwald, J. G. (1983) *Cell*, **34**, 245.
17. Tubo, R. A. and Rheinwald, J. G. (1987) *Oncogene Res.*, **1**, 407.
18. Barrandon, Y. and Green H. (1987) *Cell*, **50**, 1131.
19. Todaro, G. J. and Green H. (1963) *J. Cell. Biol.*, **17**, 229.
20. Didinsky, J. B. and Rheinwald, J. G. (1981) *J. Cell. Physiol.*, **109**, 171.
21. Cooper, D., Schermer, A. and Sun, T.-T. (1985) *Lab. Invest.*, **52**, 243.
22. Rice, R. H. and Green, H. (1979) *Cell*, **18**, 681.
23. Eckert, R. L. and Green, H. (1986) *Cell*, **46**, 583.
24. Thacher, S. M. and Rice, R. H. (1985) *Cell*, **40**, 685.
25. Peehl, D. M. and Ham, R. G. (1980) *In Vitro*, **16**, 526.
26. Tsao, M. C., Walthall, B. J. and Ham, R. G. (1982) *J. Cell. Physiol.*, **110**, 219.
27. Boyce, S. T. and Ham, R. G. (1983) *J. Invest. Dermatol.*, **81**, 33s.
28. LaRocca, P. J. and Rheinwald, J. G. (1985) *In Vitro Cell. Dev. Biol.*, **21**, 67.
29. Rheinwald, J. G., Jorgensen, J. L., Hahn, W. C., Terpstra, A. J., O'Connell, T. M. and Plummer, K. K. (1987) *J. Cell Biol.*, **104**, 263.
30. Rheinwald, J. G. and O'Connell, T. M. (1985) *Ann. N.Y. Acad. Sci.*, **21**, 259.

CHAPTER 6

Interleukin-2-driven cloned T-cell proliferation

MICHAEL B. PRYSTOWSKY

1. INTRODUCTION

The generation of an effective immune response usually involves the activation and clonal expansion of antigen-specific T-lymphocytes. Stimulation of a T-lymphocyte from a resting G_0 state requires two signals (1). The first signal involves the interaction of antigen in the appropriate form (in the context of a major histocompatibility complex molecule) with the receptor for antigen which is expressed on mature resting G_0 lymphocytes (2). Stimulation of the antigen receptor causes the expression of the receptor for interleukin-2 (IL2) and the production of lymphokines which usually include IL2 (3–5). This initial antigenic stimulation leads to increases in RNA and protein synthesis and the transition from G_0 to G_1 (6). While there are specific instances when stimulation of the antigen receptor alone will drive cell division (7), it has been found under most conditions that G_1 progression requires a second signal (e.g. IL2 or IL4). For most T-lymphocytes the second signal involves the interaction of IL2 with the high affinity IL2 receptor (8, 9). T-lymphocytes which have been stimulated recently with antigen but are no longer proliferating can be differentiated easily from resting (G_0) or naive T-lymphocytes. These activated T-cells in a postmitotic G_1 transition state (10) are characterized by the presence of IL2 receptors and the ability to proliferate in response to IL2 alone (8, 11).

Cloned T-lymphocytes possess many of the characteristics of normal activated T-lymphocytes in that they maintain their specific antigenic reactivity; they produce lymphokines in response to stimulation with antigen; and they proliferate in response to IL2 alone (12, 13). When grown *in vitro*, cloned T-cells proliferate in response to antigen plus IL2 and then return to a quiescent G_1 resting state. The maintenance of normal function and the ability to achieve a resting state distinguishes antigen-reactive cloned T-cells from cell lines that are solely dependent upon IL2 for growth. Under standard maintenance conditions, an alloreactive cloned T-cell has maintained a normal karyotype for over 8 years in culture (P. C. Nowell and M. B. Prystowsky, unpublished observation). Because the cloned T-cell system consists of a relatively homogeneous population of target cells and a pure stimulus, it provides an excellent model system to determine the events that are required for IL2-driven T-cell proliferation. The techniques given below describe how to culture murine T-lymphocytes, how to derive cloned T-cell lines, and how to analyse cells simultaneously for antigen expression and DNA content.

2. PREPARATION OF CULTURE MEDIUM

2.1 **Preparation of stock solutions**

The composition of components of the medium for culturing lymphocytes are given in *Table 1*.

(i) To prepare 100× additives, add folic acid to 200 ml of Dulbecco's phosphate-buffered saline (DPBS). While stirring add a few drops of 5 M NaOH to dissolve the folic acid. Then add the remaining nutrients and bring the volume to 250 ml with DPBS. Filter the solution through a Nalgene 500 ml 0.22 μ sterile disposable filter unit and dispense 10-ml aliquots into Falcon 2059 polypropylene tubes. Store aliquots at $-20°C$.

(ii) To prepare 100× Mops, dissolve Mops in 150 ml of distilled/deionized water (Milli-Q water can be used), adjust pH to 7.2 with 5 M NaOH and bring the volume to 250 ml. Filter and store as for additives.

(iii) To prepare 200× β-mercaptoethanol (β-ME), add 0.2 ml of β-ME to 286 ml of DPBS (this should be done in a fume hood if possible). Filter and store as for additives except store in 5-ml aliquots.

(iv) To prepare foetal calf serum (FCS), thaw one 500-ml bottle of FCS in a 37°C water bath. To remove complement and to reduce the chance of mycoplasma infection, the serum is inactivated by heating for 60 min at 56°C. Decant clear serum (usually 450 ml) into a sterile bottle and dispense into 10-, 50- or 100-ml aliquots. The dross can be discarded or saved as a source of protein for other uses.

2.2 **Culture medium**

While working in a laminar flow hood, to 1 litre of DMEM add 10 ml of 100× additives, 10 ml of 100× Mops, 10 ml of 100× Pen–Strep, 5 ml of 200× β-ME, and

Table 1. Components of medium for culturing lymphocytes.

DMEM[a]	Dulbecco's modified eagle medium with L-glutamine, with 4500 mg/litre D-glucose, without sodium pyruvate
100× Pen–strep[b]	2× penicillin–streptomycin 10000 IU/ml and 10000 μg/ml
100× Additives (for 250 ml)	150 mg folic acid
	900 mg of asparagine
	5400 mg of glutamine
	2900 mg of arginine
	2775 mg of sodium pyruvate
100× Mops (for 250 ml)	52.33 g of morpholinopropanesulphonic acid
200× β-ME	0.2 ml of β-mercaptoethanol
FCS	Selected lot of foetal calf serum

[a] Gibco; this medium contains 3.7 g/litre sodium bicarbonate and should be used with 10% CO_2. However, cloned T-cells will grow well in this medium in 5% CO_2.
[b] Flow Laboratories.

20–100 ml of FCS. Dispense into 100-ml or 250-ml aliquots and store at 4°C. Culture medium is usually used within 2 weeks of preparation. Because glutamine is labile, after 2 weeks of storage, add 1 ml of 200 mM glutamine to each 100 ml of culture medium.

3. OBTAINING A SINGLE CELL SUSPENSION OF MURINE SPLENOCYTES

3.1 Harvesting spleens

Instruments required for the harvesting of spleens are given in *Table 2*.

  (i) In the appropriate animal facility, place clean scissors and forceps in a beaker containing 95% ethanol. Light the alcohol burner.
 (ii) Kill the mice by cervical dislocation and place them on their right side.
(iii) Wet the skin covering the thorax and abdomen with 70% ethanol. Elevate the skin at the junction between the thorax and abdomen on the left side and cut it (for ~1 cm) using straight blade scissors. Just prior to using the scissors burn the 95% ethanol off.
 (iv) Using gloved hands peel the skin down over the abdomen. The spleen, a reddish brown organ about 1 cm in length, should be visible on the left side. Using the forceps and sharp-pointed scissors after burning off the alcohol, open the peritoneal cavity.
  (v) Hold the tip of the spleen in the forceps and dissect the fat and vessels from the spleen. Place the spleen in a tube containing sterile culture medium.

3.2 Preparing a single cell suspension

  (i) Loose-fitting tissue grinders are prepared by grinding a slurry of carborundum in a tissue grinder until the clearance between the two parts is increased to permit lateral movement.
 (ii) Pour the spleens and culture medium into a 15-ml loose-fitting, autoclaved, Ten Broeck tissue grinder. Grind gently until splenic fragments are no longer red. Decant the suspension into a sterile tube and wash the grinder once with 3 ml of culture medium.
(iii) Pellet the cells by centrifugation at 1000 r.p.m. (260 g) at room temperature for 10 min. Resuspend in 10 ml of fresh culture medium, repeat centrifugation and resuspension.

Table 2. Instruments for harvesting spleen cells.

Alcohol burner
Wash bottle containing 70% ethanol
Fine-pointed forceps
Straight blade scissors
Curved sharp-pointed scissors

(iv) Immediately after resuspending, centrifuge the cells at approximately 100 r.p.m. for 30 sec to remove connective tissue and other large particles. Remove the suspended cells from the particulate debris and determine the number of viable cells by trypan blue exclusion.

4. DERIVING CLONED T-LYMPHOCYTES

Cloned T-lymphocytes have been used extensively during the past decade to characterize subpopulations of T-cells functionally and biochemically. The early work, methods, potential applications and conceptual approaches to obtaining different types of cloned T-cells have been collected and presented in great detail (12). Most of the work by immunologists using cloned T-cells has been focused appropriately on the antigen-specific or functional aspects of the cells. To obtain an effective immune response involving T-cells, antigen-specific clonal expansion (i.e. proliferation) is required. Since cloned T-cells represent a type of normal T-cell which can exist in a resting state and be driven to proliferate with a purified growth factor (IL2), these cells with IL2 represent an excellent model system for studying proliferation.

Since our purpose is to study IL2-driven proliferation, the methods described below will permit the isolation of either CD8$^+$ 'cytolytic' T-cells or CD4$^+$ 'helper' T-cells. While other stimuli may affect these cells differently, we will assume that molecular mechanisms involved in IL2-driven proliferation of either type of cell will be similar and appropriate as a model for studying proliferation. The easiest method for obtaining either type of cell is the one-way mixed lymphocyte culture (MLC) which will yield alloantigen-reactive T-lymphocytes.

4.1 Mixed lymphocyte culture

In a one-way MLC, one strain serves as a stimulator and the other strain is the responder. A combination of strains that produces a good proliferative response is C57BL/6 as responders and CBA/J as stimulators.

(i) Prepare a single cell suspension of spleen cells as described above from female mice of each strain. The stimulating cells are irradiated (1400 rads).
(ii) For the primary MLC, the non-irradiated C57BL/6 responding cells (2.5×10^7) and the irradiated CBA/J stimulating cells (2.5×10^7) are incubated in 20 ml of culture medium in a 25-cm^2 tissue culture flask standing upright with a loose cap at 37°C with 10% CO_2 for 7 days.
(iii) After 7 days' culture the responding cells are harvested, pelleted, and resuspended in fresh culture medium (medium should be at room temperature or 37°C). For the secondary MLC, 5×10^6 responding cells are incubated with 2.5×10^7 freshly isolated, irradiated CBA/J splenocytes as for the primary MLC.

4.2 Isolating cloned T-lymphocytes

There are two widely used methods for isolating cloned T-cells. One involves limiting dilution; the principles and practice of this method have been described previously (14). The second method uses micromanipulation which will be described below.

(i) Responding cells from the secondary MLC are pelleted and resuspended in fresh culture medium. From 10^3 to 10^4 responding cells are then put in a 60-mm Falcon 1007 Petri dish in 5 ml of culture medium and allowed to settle.

(ii) Prepare 9 inch borosilicate Pasteur pipettes (Fisher Scientific) by plugging the end with cotton, autoclaving and, just prior to use, drawing out the end to a fine opening using a Bunsen burner. Alternatively, a capillary tube can be used. Using an inverted microscope a single cell is collected in about $20\,\mu$l and placed in a microwell (96-well culture plate) that already contains 10^6 irradiated stimulating splenocytes and 10 U/ml of recombinant IL2 (human or mouse) in a total volume of $200\,\mu$l. Potential clones are cultured for 7 days as for the MLC. Thirty to 40 wells should yield several clones.

(iii) At the end of 1 week a cluster of cells should be visible in some of the wells. Cloned T-cells often have irregular shapes and thus are not small round cells. At this time approximately $150\,\mu$l of medium should be removed without disturbing the cells. The medium should be replaced with $150\,\mu$l of culture medium containing 10^6 freshly prepared, irradiated, stimulating splenocytes and 10 U/ml of IL2. This procedure should be carried out weekly.

(iv) If the clones are growing well, then the microwells should be confluent in 2–4 weeks. At that time the cells should be transferred to 24-well culture plates.

4.3 Maintenance and expansion of cloned T-cells

(i) Cloned T-cells are maintained by weekly passage in the presence of antigen and IL2. Thus, 2×10^4 to 5×10^4 cloned T-cells, 6×10^6 irradiated stimulating splenocytes and 10 units of recombinant IL2 in about 1 ml of culture medium in a 24-well plate are incubated at 37°C in 5–10% CO_2 for 1 week. In practice, a suspension containing splenocytes and IL2 in culture medium is prepared and 1 ml is dispensed in each well of a 24-well culture plate. Then cloned T-cells are harvested, counted and 25–$100\,\mu$l of cloned T-cells are added to each 1-ml culture. The yield from these 1-ml cultures should be about 10^6 cells per well. At the end of the week the T-cells sometimes adhere weakly to the plastic. To remove T-cells, fire-polish the end of a controlled-drop borosilicate Pasteur pipette (Fisher Scientific) and scrape the bottom while pipetting.

(ii) To expand cloned T-cells, culture 10^6 cloned T-cells, 4×10^7 irradiated stimulating cells and 10 U/ml of recombinant IL2 in 10 ml of culture medium in a 60-mm Falcon 1007 Petri dish at 37°C in 10% CO_2. The yield should be about 10^7 to 2×10^7 cells per dish in 1 week. Also the cloned T-cells generally do not adhere to the plastic dishes.

4.4 General principles

(i) Variant cells will appear with time. Thus clones should be recloned every 3–4 months according to the procedure described under Section 4.2.

(ii) It is important to reclone when the cells are growing well. It has been my experience that when recloning is necessary because of poor cell growth, recloning is impossible.

(iii) For long-term growth it is important that cells return to a resting state (small cells, not blasts) at the end of the 7-day culture cycle; overstimulation can produce an unresponsive cell (15). Also, small resting cloned T-cells are larger than fresh splenocytes; the comparison is between the resting cloned T-cell and an IL2-stimulated cloned T-cell.

(iv) For some unknown reason the cultures are maintained in 24-well plates and expanded into dishes for experiments. Once cells have been expanded into dishes they are no longer passaged but only used for experiments. This may be just tradition and passage from dishes may work well but we have not tested this.

4.5 Characterization of cloned T-cells

Southern blot analysis using the β chain of the T-cell receptor as a probe can be used to determine if the lines derived are clonal. Immunofluorescence analysis with antibodies to CD4 and CD8 can be used to determine cell surface phenotype. Responsiveness to a variety of growth factors can be tested. At this time it would appear to be unnecessary to determine antigenic specificity and immunologic function if the cells are only to be used to study the regulation of IL2-driven proliferation.

5. DETERMINATION OF DNA CONTENT AND ANTIGEN EXPRESSION

One major advantage to the cloned T-cell system is that the cells grow in suspension and can be analysed easily using multiparameter flow cytometry. This method enables the determination of DNA content and specific membrane or intracellular protein expression simultaneously. This permits the correlation of gene expression with cell cycle status (16); an example of this type of analysis is given in *Figure 1* where cyclin expression and DNA content are measured simultaneously, in the cloned T-cell L2, 30 h after stimulation with 100 U/ml of purified recombinant IL2 (kindly provided by Cetus Corporation).

(i) Suspend $1-3 \times 10^6$ cells in 10 ml of DPBS and centrifuge for 10 min at 200 g (this constitutes a wash step).

(ii) Resuspend the cell pellet in 10 ml of paraformaldehyde–DPBS solution. (For the composition of this solution and others mentioned in this method see *Table 3*.) Allow to fix for 10 min and then centrifuge as above.

(iii) Resuspend the cell pellet in Triton X-100–DPBS solution. Allow to permeabilize for 3 min, then centrifuge.

(iv) Resuspend the cell pellet in an appropriate dilution of primary antibody. Allow to incubate for 1 h, then centrifuge.

(v) Wash the cells two to three times.

(vi) Resuspend the cell pellet in an appropriate dilution of fluores-

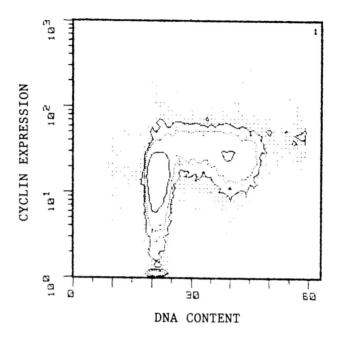

DNA CONTENT

Figure 1. Cyclin expression and DNA content in an IL2-stimulated cloned T-lymphocyte. The alloreactive cloned T-lymphocyte L2 (10^6 cells/ml) was stimulated with 100 U/ml of purified, human recombinant IL2 for 30 h. It is important to note that when passaging this cloned T-cell 10 U/ml of IL2 are used but when studying IL-2-driven proliferation 100 U/ml of IL2 are used because higher concentrations of cells are being used.

Table 3. Reagents and solutions for dual immunofluorescence.

Reagents	Cell suspension
	Dulbecco's phosphate-buffered saline (DPBS)
	Paraformaldehyde, EM grade (Polysciences)
	Triton X-100 (Sigma)
	Primary and fluoresceinisothiocyanate (FITC)-conjugated secondary antibodies
	Ribonuclease A (RNase, Worthington)
	Propidium iodide (Calbiochem)
	Bovine serum albumin (BSA, Sigma)
	Sodium azide (Sigma)
Working solutions	Paraformaldehyde–DPBS: 1% paraformaldehyde in DPBS
	Triton X-100–DPBS: 1% Triton X-100 in DPBS
	RNase–DPBS: 150 U RNase/ml of DPBS
	Propidium iodide–DPBS: 50 μg of propidium iodide/ml of DPBS
	Antibody diluent: 1 mg BSA/ml of DPBS with 0.02% sodium azide

ceinisothiocyanate (FITC)-conjugated secondary antibody. Allow to incubate for 30 min, then centrifuge.

(vii) Wash the cells two to three times.

(viii) Resuspend the cell pellet in RNase–DPBS solution. Allow to incubate for 20 min at 37°C, then centrifuge.

(ix) Resuspend the cell pellet in 1 ml of propidium iodide–DPBS solution. Allow cells to incubate for 1 h prior to flow cytometric analysis.

Unless stated otherwise, all procedures are carried out at 4°C. For storage up to 1 week, stained cell suspensions are placed in the dark at 4°C with 0.02% sodium azide.

6. IL2-DRIVEN T-CELL PROLIFERATION

Using an IL2-stimulated cloned T-cell as a model for cellular proliferation, we have shown the following.

(i) Specific changes in protein synthesis related to proliferation (17, 18).
(ii) Increases in protooncogene mRNA levels (19, 20).
(iii) An increase in potassium conductance (21) which is necessary for volume regulation (22).

Additionally, a cDNA library has been prepared using poly(A)$^+$ RNA from IL2-stimulated cloned T-cells and screening of this library has yielded several growth-regulated sequences. We anticipate that the IL2-stimulated T-cell will be an excellent model for studying the regulation of gene expression during cellular proliferation.

7. ACKNOWLEDGEMENTS

Many of the detailed procedures described in this chapter were worked out by members of Dr Frank Fitch's laboratory. Dr Charles Clevenger provided the dual immunofluorescence protocol and the data for *Figure 1*. I would like to thank Dr Frank Fitch and Dr Pierette Shipman for critically reading this chapter.

8. REFERENCES

1. Larsson, E.-L., Iscove, N. N. and Coutinho, A. (1980) *Nature,* **283**, 664.
2. Unanue, E. R., Beller, D. I., Lu, C.-Y. and Allen, P. M. (1984) *J. Immunol.,* **132**, 1.
3. Kronke, M., Leonard, W. J., Depper, J. M. and Greene, W. (1985) *J. Exp. Med.,* **161**, 1593.
4. Hemmler, M. E., Brenner, M. B., McLean, J. M. and Stominger, J. L. (1984) *Proc. Natl. Acad. Sci. USA,* **81**, 2172.
5. Prystowsky, M. B., Ely, J. M., Beller, D. I., Eisenberg, L., Goldman, J., Goldman, M., Goldwasser, E., Ihle, J., Quintas, J., Remold, H., Vogel, S. N. and Fitch, F. W. (1982) *J. Immunol.,* **129**, 2337.
6. Darzynkiewicz, Z. and Traganos, F. (1982) In *Genetic Expression in the Cell Cycle.* Padilla, G. M. and McCarty, K. S. (eds), Academic Press, New York, p. 103.
7. Moldwin, R. L., Lancki, D. W., Herold, K. C. and Fitch, F. W. (1986) *J. Exp. Med.,* **163**, 1566.
8. Cantrell, D. A. and Smith, K. A. (1984) *Science,* **224**, 1312.
9. Robb, R. J., Greene, W. J. and Rusk, C. A. (1984) *J. Exp. Med.,* **160**, 1126.
10. Richman, D. P. (1980) *J. Cell Biol.,* **85**, 459.
11. Robb, R., Munck, A. and Smith, K. (1981) *J. Exp. Med.,* **154**, 1455.
12. Fathman, C. G. and Fitch, F. W. (eds) (1982) *Isolation, Characterization, and Utilization of T Lymphocyte Clones.* Academic Press, New York.
13. Prystowsky, M. B., Otten, G., Pierce, S. K., Shay, J., Olsham, J. and Fitch, F. W. (1985) In *Lymphokines* Pick, E. (ed.), Academic Press, New York, Vol. 12, p. 13.
14. Miller, R. G. (1982) In *Isolation, Characterization, and Utilization of T Lymphocyte Clones.* Fathman, C. G. and Fitch, F. W. (eds), Academic Press, New York, p. 220.

15. Wilde, D. B., Prystowsky, M. B., Ely, J. M., Vogel, S. N., Dialynas, D. P. and Fitch, F. W. (1984) *J. Immunol.*, **133**, 636.
16. Clevenger, C. V., Bauer, K. D. and Epstein, A. L. (1985) *Cytometry*, **6**, 208.
17. Sabath, D. E., Monos, D., Lee, S., Deutsch, C. and Prystowsky, M. B. (1986) *Proc. Natl. Acad. Sci. USA*, **83**, 4739.
18. Moore, K. S., Sullivan, k., Tan, E. an d Prystowsky, M. B. (1987) *J. Biol. Chem.*, **262**, 8447.
19. Reed, J. C., Sabath, D. E., Hoover, R. G. and Prystowsky, M. B. (1985) *Mol. Cell. Biol.*, **5**, 3361.
20. Reed, J. C., Alpers, J. D., Scherle, P. A., Hoover, R. G., Nowell, P. C. and Prystowsky, M. B. (1987) *Oncogene*, **1**, 223.
21. Lee, S. C., Sabath, D., Deutsch, C. and Prystowsky, M. B. (1986) *J. Cell Biol.*, **102**, 1200.
22. Lee, S. C., Price, M., Prystowsky, M. B. and Deutsch, C. (1988) *Am. J. Physiol.*, **254**, C286.

CHAPTER 7

Muscle cell cultures

JAMES R. FLORINI, DAINA Z. EWTON, ELISSA FERRIS
and BERNARDO NADAL-GINARD

1. INTRODUCTION

1.1 Properties of muscle cell cultures

Muscle cell cultures are widely employed in studies of biological processes. Muscle is the most abundant constituent of the animal body, and it's formation and function are of importance to health and survival of the individual as well as to nutrition of the society. Conditions have been found in which a substantial portion of the properties and constituents of intact muscle can be observed in cultured cells, and the nature of muscle makes this tissue a very attractive subject for study of differentiation. In culture, skeletal muscle cells fuse to form large multinucleated myotubes which exhibit spontaneous contractions. Cardiac muscle cells, both in culture and in the heart, do not fuse, but the individual cells contract; in dense cultures, large numbers of cells contract at the same time, just as they do in the heart. These observations, coupled with a large number of physiological, biochemical, and molecular biological determinations, provide assurance that the results obtained with cultured cells are representative of many processes that occur in muscle *in vivo*.

Initially, cultured skeletal muscle cells exhibit normal proliferation when incubated under appropriate conditions. However, after some time (frequently following a change of medium to one that favours differentiation rather than proliferation of the cells) they undergo a process (1) usually called commitment to myogenic differentiation, after which they are no longer capable of DNA synthesis. There is then a closely coordinated expression of a number of muscle-specific genes, fusion of numerous individual myoblasts to form extensive myotubes, and formation of the contractile apparatus. The nature of the commitment process is not completely understood and there is some disagreement about the way it is controlled, but increased availability of purified growth factors and other medium constituents promises an improved understanding of these important processes in the near future.

1.2 Comparisons of cell lines and primary myoblast cultures

Much of the work on muscle cell growth and differentiation has been done on cell lines, particularly the L6 rat myoblast line isolated by Yaffe (2). It has been reported (3) that this line lacks at least one of the early steps (transient formation of cardiac α-actin) that occurs in the differentiation of normal skeletal muscle, and it is also true that myotubes formed by L6 cells exhibit spontaneous contractions less consistently than in primary or secondary cultures of rat, chick or quail muscle cells. Nevertheless, the substantial advantages of working with cloned, reproducible cultures have prompted both laboratories represented in this chapter to use L6 cells for initial studies, technique development, elucidation of major points, etc., and to employ primary cultures only occasionally to verify that observations with the cell lines reflect the situation in muscle *in vivo*.

Use of primary cultures (or secondary or tertiary cultures derived from them) requires the investigator to overcome two problems: (i) contamination by non-muscle cells; and (ii) lack of consistency from one preparation to another. The first is usually met by careful dissection of the muscle to minimize contamination by connective tissues and by preplating to selectively remove the more rapidly adhering 'fibroblasts' (in this context, 'fibroblasts' means all non-muscle cells). This problem can also be minimized by use of Japanese quail muscle; cultures from this source generally exhibit a much lower contamination by non-muscle cells (4).

Lack of consistency from one preparation to another is probably not as great a problem as it may seem. Use of well-characterized source animals and carefully standardized techniques eliminates a great deal of experiment-to-experiment variability, and most experimental designs depend primarily on comparisons within a single preparation of cells. Plating the cells at clonal density and doing the analyses on the basis of the clones rather that of mass cultures, as has been done extensively by Hauschka's group, can provide a great deal of reliable data from relatively few cells. Furthermore, consistency within cloned cell lines is not as great as might be assumed. It is generally observed that cell lines show a considerable 'drift' with time, and (as will be shown later) there are substantial differences between the L6E9 subline used in Boston and L6A1 clone used in Syracuse. In addition, both L6 sublines show a decrease with time in culture in fraction of cells that fuse to form myotubes, probably because there is some loss of fusing cells at each passage. Both laboratories have independently developed standard procedures of starting fresh cultures (from frozen stocks) at approximately monthly intervals in order to keep properties of the cultures as constant as possible.

2. CONDITIONS FOR PROLIFERATION AND DIFFERENTIATION OF MUSCLE CELL CULTURES

2.1 **Cell lines**

2.1.1 *L6A1 rapidly fusing subline of rat myoblasts (Syracuse)*

The culture media commonly used with L6A1 cells are given in *Table 1*.

Muscle cells are usually shipped as actively growing monolayer cultures in T25 flasks, completely filled with growth medium but may sometimes by shipped as frozen cultures in vials on dry ice.

The methods for the initial processing of L6A1 cells, their maintenance and storage are given in *Tables 2* and *3*, respectively. *Table 4* outlines measurements using L6A1 cultures.

Table 1. Media used with L6A1 cells.

Growth medium

10%[a] horse serum (HS)
 2% chick embryo extract (CEE) (Gibco)
 1% antibiotic–antimycotic solution (100×) (Sigma)

in Dulbecco's Modified Eagle Medium (DMEM) (with 1.0 g D-glucose/litre) (Gibco)

Fusion/differentiation medium

0.3% μM insulin (Sigma)
0.05% bovine serum albumin (BSA) (Sigma) or 1% HS
1% antibiotic–antimycotic solution

in DMEM

Note: first dissolve insulin in 0.01 M HCl (~0.4 ml) and then dilute with DMEM plus BSA to 3×10^{-7} M. All BSA batches should be checked for the presence of mitogens; we found one that was so active that we concluded that it contained as much as 5% bovine somatomedin.

Freezing medium

14% HS
10% dimethylsulphoxide (DMSO)
 1% antibiotic–antimycotic solution

in DMEM

[a] All % concentrations are v/v.

Table 2. Initial processing of L6A1 cells.

A. *Monolayer cultures*

1. Upon arrival, examine the cultures using a phase-contrast microscope, to be sure that fusion of myoblasts into myotubes has not begun. If some myotubes are present, subculture the cells immediately, to prevent loss of more cells to differentiation. Otherwise, incubate the cultures overnight at 37°C.
2. The next day, remove the growth medium, either by pouring it out or by aspirating using a sterile Pasteur pipette connected to a trap and vacuum pump.
3. Add 4 ml of 0.25% trypsin (Gibco) and proceed to establish new stock cultures as detailed in *Table 3*.

B. *Frozen cultures*

1. Upon arrival, immediately place the frozen culture in liquid N_2 or in a -70°C freezer.
2. When needed, thaw the cultures as described in *Table 3C*.

Table 3. Maintenance and storage of L6A1 cells.

A. *Maintenance of stock cultures*

1. Remove medium from established stock cultures [~1–2 × 10⁶ cells per 100 mm (Corning) tissue culture dish].
2. Add 4ml of 0.25% trypsin (Gibco).
3. Incubate for 5 min at 37°C. Examine under microscope (phase-contrast) to be certain cells have lifted.
4. Pull solution in and out of Pasteur pipette gently 3–4 times to dissociate clumps and make certain all cells are removed from the plate. Direct output of the pipette across surface of cells layer.
5. Transfer to a 50-ml centrifuge tube containing 5–10 ml of DMEM containing 10% HS to suppress further trypsin action.
6. Centrifuge for 5 min at about 500 g. (This is setting 6 on an International IEC Model CL Centrifuge.)
7. Pour off the supernate.
8. Suspend the cells in 2 ml of growth medium, using a Pasteur pipette, then dilute to 5–10 ml.
9. Count in a haemocytometer.
10. Plate new stock plates at 1–2 × 10⁵ cells/8 ml of growth medium in each 100-mm tissue culture dish.

Note: we normally establish new stock plates on Monday and Friday and feed the cultures with fresh growth medium on Wednesday. If fresh medium is not added within 48–72 h, cells are likely to fuse and thus be lost to further proliferation.

B. *Freezing cells*

1. Grow up a large amount of stock as soon as possible, as outlined above.

Note: it is best to freeze L6 myoblasts within a week (one to two passages) of receiving a new shipment of cells. One confluent stock plate containing ~2 × 10⁶ cells can be split into 20 stock plates, which in a week should yield enough cells to make 40 vials.

2. Trypsinize and centrifuge as described in A(1–7). Suspend in cold freezing medium at about 10⁶ cells/ml.
3. Aliquot into freezer vials, about 10⁶ cells/litre per ml vial.
4. Freeze first at −20°C for 2 h, then transfer to a −70°C freezer or the vapour phase of liquid N₂. Under these conditions, the cells retain >90% viability for at least 1 year.
5. Fresh cells should be removed from the freezer every 3–4 weeks to maintain consistent growth and differentiation characteristics. Establish fresh cultures as outlined below under thawing cells.

C. *Thawing cells*

1. Thaw quickly by placing the vial in a 37°C water bath.
2. Transfer the cell suspension to a 50-ml centrifuge tube with about 10 ml growth medium.
3. Centrifuge for 4 min at 500 g.
4. Resuspend the cells in growth medium and plate at ~3 × 10⁵ cells/100 mm.
5. The next day, wash the cultures with DMEM and add fresh growth medium.

Table 4. Measurements using L6A1 cultures.

A. *Stimulation of differentiation*

1. Plate the experimental cultures at about 1.2×10^5 cells/8-cm^2 dish in 2 ml of medium containing 5% HS and 1% CEE in DMEM; incubate overnight at 37°C in a CO_2 incubator.
2. Aspirate the medium and wash the cultures with 2 ml of serum-free DMEM for 15 min at 37°C.
3. Add 2 ml of fusion/differentiation medium.

Note: if serum-free conditions are necessary, 0.3 μM insulin is added in the presence of DMEM containing 0.05% BSA. However, under these conditions fusion of myoblasts into myotubes is accompanied by some of cells lifting. Better differentiation is obtained if insulin is added in the presence of 1% HS in DMEM. Most cells stay attached and more extensive myotube formation occurs.

4. Measure the extent of differentiation after 72 h by percentage fusion or elevation of creatine kinase (CK).

B. *Percentage fusion*

1. Wash the cultures twice with 2 ml of PBS at room temperature.
2. Fix the cultures in absolute methanol.
    (i) Add methanol for 1 min.
    (ii) Aspirate.
    (iii) Add fresh methanol for 5 min.

Note: if staining is not being done immediately, the cultures can remain in methanol until all experimental dishes are fixed if care is taken to keep the cells covered with methanol.

3. Aspirate off the methanol.
4. Add approximately 2 ml of Wright's stain: 5–10 min.
    Wright's stain   0.3 g
    methanol   100 ml
    glycerol   3 ml
5. Pour off the Wright's stain.
6. Add 2 ml of Giemsa stain (dilute first 1 : 10 in distilled water); leave on for 15 min.
7. Flush with water.
8. Let dry upside down.
9. Observe under a microscope, using a 40× objective.
10. Count the percentage of nuclei in myotubes.
    (i) Count at least 10 fields per dish.
    (ii) Count approximately 1000 nuclei per dish.

Note: cells are considered to be myotubes if they contain three or more nuclei.

C. *Creatine kinase* (modification of ref. 4a)

1. Wash the cell monolayer two times with cold PBS.
2. Add 250 μl of 0.05 M glycylglycine, pH 6.75.

continued

Table 4. (*continued*)

C. *Creatine kinase* (modification of ref. 4a)

3. Stack up approximately eight dishes, tape them together and place in a freezer ($-20°C$ for up to 1 week, or $-70°C$ for up to 1 month). Take care that plates are not tipped until the buffer is frozen.
4. Thaw the cultures by placing plates on top of a layer of crushed ice.
5. Scrape with rubber policeman, rotating the culture dish to be sure all the cells are detached from the surface.
6. Transfer the cell lysate to test tubes; keep on ice.
7. When all the dishes have been processed, vortex; aliquot 175 μl to glass tubes for a DNA assay; aliquot 33 μl to test tubes for a CK assay.
8. Add 1 ml of prewarmed (30°C) CK reagent:
   - 20 mM glucose
   - 10 mM Mg acetate
   - 1 mM ADP
   - 10 mM AMP
   - 0.8 mM NADP
   - 20 mM creatine phosphate
   - 1 U/ml hexokinase
   - 0.5 U/ml glucose-6-phosphate dehydrogenase
   in 0.1 M glycylglycine, pH 6.75 (add dithiothreitol, 1.54 mg/ml, just before use).
9. Read the change in absorbance at A_{340} during a 5-min incubation. This is done most conveniently with a recorder attached to a spectrophotometer giving linear A_{340} output.
10. Calculate mU/ml.

(One unit = amount of CK that converts 1 μmol of NADP to NADPH in 1 min.)

2.1.2 $L_6 E_9$ subline of rat myoblasts (Boston)

Solutions commonly used with $L_6 E_9$ cells are given in *Table 5*. Methods for the initial processing of $L_6 E_9$ cells, their maintenance and storage are outlined in *Tables 6* and *7*, respectively. Measurements using $L_6 E_9$ cells are given in *Table 8*.

2.1.3 *C2 mouse myoblasts*

Solutions commonly used with C2 mouse myoblasts are given in *Table 9*. The maintenance of stock cultures and stimulation of differentiation are given in *Tables 10* and *11*, respectively.

Table 5. Solutions used with L_6E_9 cells.

Growth medium

20% foetal bovine serum (FBS)
 1% Penicillin–streptomycin solution (Gibco)

      in DMEM with 4.5 g of D-glucose/litre (Gibco)

Fusion medium

 5% HS
 1% Penicillin–streptomycin solution

      in DMEM

PBS and versene

20%	EDTA	1 ml
0.6%	Phenol Red	1 ml
	10× PBS	100 ml

      dilute to 1 litre with water

Pancreatin solution

10× pancreatin 4× N.F. (Gibco)	1 ml
PBS and versene	9 ml

Freezing medium

20% FBS
10% DMSO

      in DMEM

Table 6. Initial processing of L_6E_9 cell lines.

A. *Monolayer cultures*

Process the cells essentially as described in *Table 2* for L6A1 cells, except using the different solutions for subculturing as described below.

B. *Frozen cultures*

Process the cells as described for L6A1 cells in *Table 2*.

Table 7. Maintenance and storage of L_6E_9 cells.

1. Remove the spent medium from a parent 100-mm diameter plate (~2.5–5 million cells/Falcon no. 3003 tissue culture dish).
2. Wash the cells with 5 ml of PBS and versene.
3. Apply 1 ml of pancreatin solution.
4. Incubate for 5 min at room temperature.
5. Tap the plate until most of the cells appear to be detached from the plate surface.
6. Add 4 ml of growth medium.
7. Disperse the cells by pipetting gently until a single cell suspension is evident.
8. Aliquot 0.5 ml of the cell suspension directly to 10 ml of growth medium in each Falcon no. 3003 tissue culture dish. We subculture our cells every 48–72 h without feeding; in this way we maintain a homogeneous population up to 6 weeks.
9. Count the parent plate in a haemocytometer to verify initial cell inoculum.

Freezing and thawing L_6E_9 cells

Cells are frozen and thawed as described for L6A1 cells in *Table 3*, except that the centrifugation upon thawing is omitted.

Note: the DMSO is diluted to 1% when 1 ml of a frozen suspension is diluted to 10 ml of growth medium. We find good viability in the presence of DMSO at this concentration.

Table 8. Measurements using L_6E_9 cultures.

Stimulation of differentiation

1. Plate the cells at 2.5–5 × 10^5 cells per 100-mm diameter plate.
2. Aspirate the growth medium 40 h after seeding and replace with differentiation medium.
3. Observe myotube formation 48–72 h later.

Myotubes are not well maintained beyond 72 h using this protocol. More stable myotubes may be obtained by using the following protocol, although myotube formation takes longer.

1. Seed cells in growth medium at 2.5–5 × 10^5 cells per 100-mm diameter plate.
2. Aspirate the growth medium 24 h later and replace with differentiation medium.
3. Observe myotube formation 6 days later. Myotubes may be maintained for 2 weeks without feeding under these conditions.

Table 9. Solutions used with C2 mouse myoblasts.

Growth medium

20%   FBS
0.5% CEE
1%   antibiotic–antimycotic solution in DMEM

Fusion/differentiation medium

A. *Slow* differentiation   or   B. *Rapid* differentiation

5–10% HS   0.3 μM insulin
  1% antibiotic–antimycotic solution   0.05% BSA or 1% HS
  in DMEM   1% antibiotic–antimycotic solution
                                                                        in DMEM

continued

Table 9. *continued*

0.05% Trypsin–EDTA

0.02% EDTA in PBS 16 ml
0.25% trypsin   4 ml

Freezing medium

10%   FBS
10%   DMSO
 1%   antibiotic–antimycotic solution
in DMEM

Table 10. Maintenance of stock cultures.

1. Remove medium from established stock cultures (55-cm^2 dishes).
2. Wash the cells with 4 ml of EDTA–PBS.
3. Add 4 ml of 0.05% trypsin–EDTA.
4. Incubate for about 1 min at 37°C until the cells have lifted.
5. Dissociate clumps of cells using a Pasteur pipette.
6. Transfer to a centrifuge tube containing 5 ml of fusion medium.
7. Centrifuge for 5 min at 500 *g*; pour off supernate.
8. Suspend the cells in 5–10 ml of growth medium.
9. Count in a haemocytometer.
10. Plate new stock plates at 1×10^5 cells/55 cm^2/8ml of growth medium.

Table 11. Stimulation of differentiation.

1. Plate experimental cultures at about 1.2×10^5 cells/8 cm^2/2 ml of growth medium. Incubate overnight at 37°C.
2. Aspirate the medium and wash the culture with 2 ml of serum-free DMEM for 15 min at 37°C.
3. Add 2 ml of differentiation medium A or B.

   Note: addition of differentiation medium A will be followed by relatively slow fusion of myoblasts into myotubes over several days. Much more rapid differentiation is observed when using Medium B. Maximum activity of CK (expressed as U/mg of DNA) is reached 48 h after addition of insulin.

2.1.4 *Other myoblast cell lines*

BC$_3$H$_1$ is a myogenic cell line isolated from mouse neural tissue. It does not undergo irreversible terminal differentiation—myoblasts do not fuse to form myotubes, and CK levels are elevated at high cell density or in low FBS, but are lowered upon the incubation in 20% FBS.

Growth medium
20% FBS
 1% antibiotic–antimycotic in DMEM

Differentiation medium
1% FBS
1% antibiotic–antimycotic in DMEM

2.2 **Primary muscle cells**

2.2.1 *Primary quail myoblasts*

The preparation of cultures of primary quail myoblasts is modified from the procedure of Konigsberg (4). Solutions and equipment are outlined in *Table 12*. The preparation of cultures is detailed in *Table 13*.

Table 12. Equipment and solutions for the culture of primary quail myoblasts.

Materials and equipment

Quail eggs (*Coturnix coturnix japonica*), incubated for 9–10 days
DMEM
CEE
HS
Gauze-covered beaker (sterile)
Sterile forceps and scissors
Sterile Petri dishes
Gelatin
Tissue culture plates—gelatin coated
Sterile water
Trypsin (0.25% stock)
Hanks' balanced salt solution (BSS) or sterile PBS

Growth medium

15%   HS
10%   CEE
 1%   antibiotic–antimycotic solution
in DMEM

Note: it may be necessary to add Fungizone (Gibco) ($1–2\ \mu g/ml$) if contamination is a problem.

Trypsin, 0.05%

Dilute stock trypsin (0.25%) with Hanks' BSS (Flow Labs) or sterile PBS
  trypsin stock   4 ml
  Hanks' BSS   16 ml

Gelatin-coated culture plates

1. Wash 100 mg of gelatin several times with ice cold sterile water, centrifuging each time for 1 min at 500 *g* then pouring off the supernatant.
2. Dissolve in 500 ml of distilled water.
3. Autoclave and store in the cold room.
4. When required
   (i) flood the tissue culture plates with the gelatin solution;
   (ii) leave the plates in the incubator overnight;
   (iii) aspirate off the liquid the next morning;
   (iv) store the plates in the hood until ready to use.

2.2.2 *Primary chick myoblasts*

Primary chick myoblasts can be prepared from 11-day-old quail chick embryos by the same method used for primary quail monoblasts (*Table 13*) but substituting the following for growth medium

   15% HS
   5% CEE
   1% antibiotic–antimycotic solution
in DMEM

Table 13. Preparation of cultures.

Note: only the amount of tissue (~four embryos) which can be processed within an hour should be taken. All procedures should be done in a laminar hood under sterile conditions.

1. Rinse the eggs with alcohol.
2. Using sterile scissors, cut a hole in the egg. Using curved forceps, remove the embryos from the eggs.
3. Place the embryos (9- to 10-day-old) in a sterile Petri dish containing sterile PBS or Hanks' BSS; remove the heads.
4. One at a time, transfer the embryos to a fresh Petri dish with PBS. After removing the skin, transfer the pectoral muscle to a Petri dish with a few drops of PBS.
5. Mince the pooled muscle with curved scissors.
6. Transfer the mince with a pipette to a centrifuge tube containing 5 ml of diluted trypsin (0.05%).
7. Pipette up and down gently for 8–10 min, maintaining the temperature at 37°C—use a large bore pipette.
8. Add 5 ml of growth medium to inactivate the trypsin.
9. Filter through a sterile gauze-covered beaker—wash with an additional 5 ml of medium.
10. Centrifuge at 500 g.
11. Remove the supernate and disperse the pellet in 5 ml of growth medium; after pipetting, add an additional 5–10 ml of medium.
12. Count the number of cells using a haemocytometer.
13. Plate in gelatin-coated tissue culture dishes at 10^7 cells per 100-mm dish.
14. After the cells have attached (2–4 h), gently wash the plates with PBS and then add fresh growth medium.

Secondary cultures

1. The next day, wash the primary cultures with PBS.
2. Lightly trypsinize the plates, one at a time (~2 ml trypsin per 100-mm plate; it is best to monitor the plates under the microscope occasionally while trypsinizing). When about half the cells have rounded up, add growth medium to the plate and gently pipette, transferring the detached cells to a centrifuge tube.

Note: this procedure takes advantage of the lower affinity of myogenic cells for the plastic surface of the tissue culture plate.

3. Count the cells using a haemocytometer.
4. Plate the cells at 2–5 × 10^4 in gelatin-coated 8-cm^2 tissue culture plates in growth medium.

2.2.3 *Primary rat myoblasts*

Details of the solutions and equipment used in the culture of primary rat myoblasts are given in *Table 14*. The culture procedure of primary rat myoblasts is described in *Table 15*.

Table 14. Solutions and equipment for primary rat myoblast culture.

Trypsin (0.25%)
Earle's BSS
Plating medium (see below)
Growth medium (see below)
Gauze-covered beaker
Watchglass
2 Petri dishes (100 mm)
Beaker with 70% ethanol
2 Straight forceps
2 Straight scissors
4 Curved forceps
1 Curved scissors
1 Trypsinizing flask with magnetic stir bar
1 Magnetic stirrer
Gelatin-coated culture plates—prepared as for primary quail myoblasts (*Table 12*)

Plating medium

10%   HS
 1%   CEE
 1%   antibiotic–antimycotic solution

in DMEM (Gibco)

Growth medium

20%   FBS
 5%   CEE
 1%   antibiotic–antimycotic solution

in DMEM

Table 15. Culture of primary rat myoblasts.

Day 1

The preparation of primary rat myoblasts

1. Cut off the heads of 1- to 3-day-old rats using scissors and rinse in cold running water at sink.
2. In the sterile hood, dip each rat in 70% ethanol, place on the watchglass, and cut off the hind legs above the thigh using scissors.
3. Place the legs in the first Petri dish containing Earle's BSS.
4. Repeat with the remaining rats.
5. Remove the skin and feet and place the thighs in a second Petri dish containing Earle's BSS.
6. Remove muscle from bones on an inverted Petri dish cover and move it to an area on the cover where a few drops of Earle's BSS have been added (use dissecting microscope) in the hood. (The covers have lower rims and are more convenient for dissecting than are the Petri dishes.)
7. Mince by cutting with curved scissors.
8. Trypsinize at room temperature for 10 min in 10 ml of 0.25% trypsin in a trypsinizing flask with gentle stirring.
9. Tilt flask ~1 min until the muscle settles to the bottom and gently pour off most of the solution and discard.
10. Add 15 ml of trypsin to the flask and gently stir for 15 min.
11. Tilt the flask and gently pour off the solution into a beaker containing 35 ml of DMEM + 15 ml of HS.
12. Repeat steps 10 and 11 four more times, pouring the trypsin into the same beaker containing DMEM + HS.
13. Pour through a gauze-covered beaker.
14. Centrifuge at 500 g for 5 min.
15. Resuspend cells in plating medium and count the cells.
16. Plate out cells at about 5×10^6 cells per 100-mm gel-coated dish in 8 ml of plating medium.

Day 2

1. Replace media with growth medium.

Day 3

1. Trypsinize the cells and transfer to a centrifuge tube.
2. Add an equal amount of plating medium and centrifuge the cells at 500 g for 5 min.
3. Resuspend the cells in plating medium and count.
4. Plate out the cells in *non*-gelatin-coated 100-mm dishes at about 6×10^6 cells per dish in 8 ml of plating medium and allow non-muscle cells to attach at 37°C for 20–30 min.
5. Gently remove media containing non-attached muscle cells.

Note: this method takes advantage of the more rapid attachment of 'fibroblasts' to give cultures substantially enriched in myogenic cells.

6. Count the cells and set up secondary muscle cells according to the experimental design in gelatin-coated dishes.

117

3. MODIFIED OR SPECIALIZED MEDIA FOR THE GROWTH OF MUSCLE CELLS

3.1 Serum-free media

In a recent review, one of us (5) tabulated the constituents of six serum-free media that had been developed for various muscle cell systems ranging from 'wild type' L6 cells to human myoblasts. The compositions of the media varied widely, from three components (6) to 12 (7) added to the basal salts/nutrient medium—which was not the same in any of the six media. In spite of this diversity, certain similarities are apparent. All of the media contain insulin at high (micromolar) levels that would allow extensive cross-reaction with the insulin-like growth factor-I (IGF-I) receptor. In these situations, insulin can probably be regarded as a relatively inexpensive substitute for IGF-I, which is (at physiological concentrations) a potent mitogen for muscle cells (8). Most of the media contain another mitogen, fibroblast growth factor (FGF) (9), which is a potent stimulator of muscle cell growth (and inhibitor of differentiation), but L6 cells do not respond to this growth factor.

In spite of the catabolic action of glucocorticoids *in vivo*, all of the media include these hormones, usually in the form of the relatively slowly metabolized dexamethasone. It seems likely that these hormones are required for normal metabolism of the cultured cells; they are components of virtually all serum-free media that have been devised for non-muscle cells.

Fetuin, which is the primary protein in FBS, is present at mg/ml levels in several of the media. This concentration raises questions of possible contamination by various growth factors, most of which exhibit full activity at the ng/ml level—that is a few parts per million contamination of the fetuin might account for the observed activity. However, in spite of various published and unpublished attempts to demonstrate an active contaminant that accounts for the activity of fetuin in this system, no such contaminant has yet replaced fetuin in the systems in which it is used. In their study of rat satellite cells, Allen *et al.* (9) found that fetuin could not be replaced by either fibronectin (10) or α_2-macroglobulin (11), both of which have been suggested as accounting for the activity of fetuin in such media.

Overall, the diversity of the media suggests that serum-free formulations are a rather individual matter, and their composition should probably be tailored for each situation for which they might be used. The summary presented by Florini (5) can provide a useful starting list of components that might be active in any specific situation.

3.2 Use of hormones and growth factors

Recombinant DNA technology has now been applied to the production of a number of hormones and growth factors that were previously available only in very small quantities, if at all. In addition to their potential usefulness in development of serum-free media (above), they can also be employed as potent controllers of important processes in muscle cells. Linkhart's group (12, 13) has clearly demonstrated that FGF is a potent inhibitor of myogenic differentiation, and Lathrop and his colleagues (14, 15) have shown that FGF blocks differentiation even in the non-fusing BC_3H_1 line. The two laboratories represented in this chapter independently reported that transforming growth factor-β (TGF-β) is a very potent but reversible inhibitor of L6 myoblast differentiation (16, 17), and Olson *et al.* (18) reported similar observations with C2 and BC_3H_1 cells at about the same time.

Other growth factors (the insulin-like family, including insulin) are strong stimulators of muscle cell growth and differentiation. The stimulation of differentiation is strikingly biphasic (18); at higher concentrations (above physiological levels) the stimulation decreases until there is essentially no difference from controls at hormone concentrations 50–100 times those giving maximal stimulation. These effects at high concentrations, coupled with the early studies with FGF, may account for the widespread but inaccurate belief that all mitogens are inhibitors of muscle cell differentiation.

3.3 Minimizing consumption of foetal bovine serum

The high cost of FBS has prompted us to investigate the possibility of using information gained in studies on serum-free media and growth factors to reduce or eliminate the need for this component of most plating and growth media for muscle cells. (Obviously this is a lesser concern in experiments on terminal differentiation.) Both laboratories have done several studies in this area, and we have come to the following conclusions.

(i) There are substantial variations among sublines so that the suggestions made here must be regarded simply as starting points that may or may not prove suitable for the cells and conditions being used in another laboratory.

(ii) In many instances, FBS can be replaced by 20% HS and very high levels of insulin (10 μg/ml). This approach takes advantage of the fact that the concentration dependency curve for the stimulation of differentiation by insulin is strikingly biphasic (19), and such a high level of insulin prevents differentiation in stock cultures of myoblasts. Under these conditions—as in all work with stock myoblast cultures—care must be taken to do medium changes and passing at sufficiently short intervals that the most active myogenic cells are not lost from the proliferative pool as a result of fusion to form postmitotic myotubes.

Obviously, this approach has substantial limitations if the experimental approach involves effects of insulin or the insulin-like growth factors. Furthermore, there have been some indications of inconsistent results with different insulin preparations, so care must be taken that the specific material being used gives the expected results before any extensive commitment is made to this approach.

(iii) Although the more recently developed semisynthetic media such as NuSerum (Collaborative Research) support rapid growth of myoblasts (C2 cells proliferated in 10% NuSerum plus 5% FBS as rapidly as in 20% FBS), they do not suppress loss of proliferative cells to differentiation. Conceivably, the component missing from NuSerum may be TGF-β, and we are currently investigating the possibility that appropriate supplementation will allow maintenance of myoblast stock cultures in proliferative states.

4. ACKNOWLEDGEMENTS

The techniques described here were developed or modified with support from USPHS grants HL11551 and AG05557, and grants from the Muscular Dystrophy Association. We are grateful for technical assistance from Ms Susan Falen, Lilie Welych and Carri Jones.

5. REFERENCES

1. Nadal-Ginard, B. (1978) *Cell*, **15**, 855.
2. Yaffe, D. (1968) *Proc. Natl. Acad. Sci. USA*, **61**, 477.
3. Hickey, R., Skoultchi, A., Gunning, P. and Kedes, L. (1986) *Mol. Cell. Biol.*, **6**, 3287.
4. Konigsberg, I. R. (1979) In *Methods in Enzymology*. Jacoby, W. B. and Pastan, I. H. (eds), Academic Press, New York, Vol. 58, p. 511.
4a. Shainberg, A., Yagil, G. and Yaffe, D. (1971) *Dev. Biol.*, **25**, 5.
5. Florini, J. R. (1987) *Muscle Nerve*, **7**, 577.
6. Florini, J. R. and Roberts, S. B. (1979) *In Vitro*, **15**, 938.
7. Askanas, V., Cave, S., Gallez-Hawkins, G. and Engel, W. K. (1985) *Neurosci. Letts*, **61**, 213.
8. Ewton, D. Z., Falen, S. L. and Florini, J. R. (1987) *Endocrinology*, **120**, 115.
9. Allen, R. E., Dodson, M. V., Luiten, L. S. and Boxhorn, L. K. (1985) *In Vitro*, **21**, 636.
10. Dollenmeier, P., Turner, D. C. and Eppenberger, H. M. (1981) *Cell. Res.*, **135**, 47.
11. Salomon, D. S., Bano, M., Smith, K. B. and Kidwell, W. R. (1982) *J. Biol. Chem.*, **257**, 1409.
12. Linkhart, T. A., Clegg, C. H. and Hauschka, S. D. (1981) *Dev. Biol.*, **86**, 19.
13. Linkhart, T. A., Clegg, C. H., Lim, R. W., Merrill, G. F., Chamberlain, J. S. and Hauschka, S. D. (1982) In *Molecular and Cellular Control of Muscle Development*. Person, M. L. and Epstein, H. F. (eds) Cold Springs Harbor Conference, p. 877.
14. Lathrop, B., Olson, E. and Glaser, L. (1985) *J. Cell. Biol.*, **100**, 1540.
15. Olson, E. N., Caldwell, K. L., Gordon, J. I. and Glaser, L. (1983) *J. Biol. Chem.*, **258**, 2644.
16. Massague, J., Cheifetz, S., Endo, T. and Nadal-Ginard, B. (1986) *Proc. Natl. Acad. Sci. USA*, **83**, 8206.
17. Florini, J. R., Roberts, A. B., Ewton, D. Z., Falen, S. B., Flanders, K. C. and Sporn, M. B. (1986) *J. Biol. Chem.*, **261**, 16509.
18. Olson, E. N., Sternberg, E., Hu, J. S., Spizz, G. and Wilcox, C. (1986) *J. Cell. Biol.*, **103**, 1799.
19. Florini, J. R., Ewton, D. Z., Falen, S. L. and Van Wyk, J. J. (1986) *Am. J. Physiol. (Cell Physiol.)*, **250**, 771.

CHAPTER 8

Cell culture of human diploid fibroblasts in serum-containing medium and serum-free chemically defined medium

VINCENT J. CRISTOFALO and PAUL D. PHILLIPS

1. INTRODUCTION

Normal human diploid fibroblast-like cells are widely used for studies of *in vitro* cell aging and in studies of growth factor action and cell proliferation. Because of their special growth characteristics the manipulation of these cells requires special attention and understanding to insure the vigorous and optimum growth of the cultures.

Cells derived from normal human tissue, explanted and grown in tissue culture, have the following characteristics: (i) the cultures possess a finite and predictable proliferative capacity most accurately measured as cumulative population doublings; (ii) the finite proliferative capacity of a cell line is inversely proportional to the age of the donor; (iii) normal human fibroblast-like cells have never been shown to spontaneously transform in culture, that is, they have never been shown to spontaneously acquire the indefinite proliferative capacity characteristic of tumour-derived or otherwise transformed cells; (iv) normal human cells can be transformed into cultures with an indefinite proliferative capacity by infection with DNA tumour viruses such as simian virus 40 (SV40) (1). There are very few reports of transformation by chemical mutagenesis (2).

For routine purposes, the serial propagation of normal human cells is most readily accomplished using serum-supplemented medium. Although this is generally acceptable for simply maintaining stock cultures there are associated problems. The use of serum in specific experimental designs can be a source of obvious as well as not so obvious complications. One must realize that serum is an extraordinarily complex biologic fluid composed of literally hundreds of bioactive components whose effects on any particular cellular response are, at best, not totally understood. Because of the vast number of known and unknown factors in serum it is obvious that a defined mitogen supplement would be preferred. There have been reports of the serial propagation of normal human fibroblasts in completely serum-free (i.e. free of undefined biological fluids) growth-factor-supplemented medium (3). However, many normal human cell lines require complex and difficult to prepare lipid supplements in addition to various growth factors to insure cell viability and proliferation following subcultivation which includes exposure to the proteolytic enzymes. To avoid many of these problems we have developed a system which supports multiple rounds of cell proliferation in a serum-free medium supplemented with various growth factors (4, 5). In this chapter we describe our procedures for maintaining cultures of normal human fibroblasts throughout their *in vitro* lifespan in serum-supplemented medium and our procedures for taking stock cultures of these cells and subcultivating them into a serum-free medium for studies of cell aging and growth factor action.

2. PROPAGATION OF CELLS IN SERUM-SUPPLEMENTED MEDIUM

The finite replicative lifespan of normal human cells is characterized by a period of rapid proliferation followed by a decline in the rate of proliferation, after which the cultures can no longer be propagated (6). This decline in proliferative capacity is interpreted as an expression of aging at the cellular level (7). Proliferative rate and the stage in the life history of a culture can be monitored best by well-controlled and reproducible cell culture procedures (8, 9). This section describes the standard subcultivation procedures we use for the study of aging in normal human diploid fibroblast-like cells maintained in serum-supplemented medium.

2.1 **Materials**

2.1.1 *Growth medium and trypsinizing medium*

We use Auto-Pow (Flow Laboratories, 7655 Old Springhouse Rd, McLean, VA 22102, USA), which is an autoclavable formulation of Eagle's modified minimal essential medium (MEM) with Earle's salts, without glutamine and without sodium bicarbonate. To this we add basal medium Eagle (BME) vitamins (Flow Laboratories). After autoclaving the medium is supplemented with sodium bicarbonate (Sigma), L-glutamine (Sigma Chemical Co., PO Box 14508, St Louis, MO 63178, USA), and foetal bovine serum (FBS) (Flow Laboratories). The medium is prepared in two stages (see *Table 1*).

In order to subcultivate the cells they must be removed from the vessel in which they are growing. For this process we use a Ca^{2+}, Mg^{2+}-free MEM which is supplemented with trypsin (*Table 2*).

2.2 **Trypsinization and harvesting of the cells**

Pour off the medium from the cell culture flask into a sterile beaker by decanting the fluid from the side opposite the cell growth surface. Alternatively, the medium can be aspirated out of the flask using a Pasteur pipette and into a vacuum flask. Rinse the cell sheet twice with 4 ml each of the trypsinizing solution. Remove the rinse solution by decanting or aspiration. Add enough trypsinizing solution to just cover the cell sheet. For example, 2 ml is sufficient for a T-75 flask. Secure the cap of the culture flask and incubate the flask for approximately 15 min at 37°C. Check periodically to see if the cells are detaching. The cells will round up as they lift off the growth surface. The process may be speeded up by gently tapping the sides of the flask. Take care not to splatter the cells against the top and sides of the flask because this will lead to errors in determining the number of cells in the flask.

Table 1. Preparation of medium for culture in serum-supplemented medium.

A.

1. To make 1 litre of medium dissolve 9.4 g of Auto-Pow powder in 854 ml of deionized H_2O and add 10 ml of the 100× stock solution of BME vitamins.
2. Mix this thoroughly.
3. Dispense 432 ml of the incomplete medium into 1-litre bottles. Screw the caps on loosely, apply autoclave indicator tape and autoclave on fast exhaust for 15 min at 121°C.
4. When the sterilization cycle is complete quickly remove the bottles from the autoclave because prolonged exposure to high temperatures destroys some of the medium components.
5. With the caps still loose place the bottles in the laminar flow hood and allow them to cool to room temperature.
6. When the cooling is complete, tighten the caps and store the medium at 4°C in the dark.

B.

The second stage in the process is carried out when the medium is ready to be used. Each 1-litre bottle contains 432 ml of sterile but incomplete medium.

1. To each bottle add 13 ml of a filter-sterilized solution of 7.5% (w/v) sodium bicarbonate, 5 ml of a filter-sterilized solution of 200 mM L-glutamine and 50 ml of sterile FBS.

This complete medium should be freshly prepared for each use.

Table 2. Preparation of Ca^{2+}, Mg^{2+}-free MEM.

1. To 900 ml of deionized H_2O add the following in the order listed: 6.8 g of NaCl, 0.4 g of KCl, 0.14 g of $NaH_2PO_4 \cdot H_2O$, 1 g of glucose (all from Sigma), 20 ml of 50× MEM amino acids, 10 ml of 100× MEM vitamins (both from Flow Laboratories), and 10 ml of a 0.5% (w/v) solution of Phenol Red.
2. Bring this to 1 litre with deionized H_2O and filter-sterilize, using a 0.2-μm pore size, under house vacuum.

At the time it is to be used this solution is supplemented as follows.

3. To 40 ml of the sterile Ca^{2+}, Mg^{2+}-free MEM add 5 ml sterile 7.5% sodium bicarbonate, and 5 ml of a sterile 2.5% trypsin solution (Flow Laboratories).
4. This should be freshly prepared and kept on ice.

When all the cells have detached from the growth surface, add enough complete medium (10% FBS-supplemented medium) to wash down the surface of the culture vessel. For a T75 flask we add 8 ml for a final harvest volume of 10 ml. Before seeding the new flasks the cell clumps must be broken up by drawing up the entire suspension into a 10-ml pipette and then allowing it to flow out gently against the wall of the vessel. Repeat the process at least three times. Repeat the same procedure using a 5-ml pipette to obtain a single-cell suspension. Until the procedure becomes routine, check the suspension under the microscope. Suspensions from several T75 flasks can be pooled into one flask. Keep the cells on ice during such procedures to inhibit cell reattachment.

Using sterile procedures, remove an aliquot from the suspension and dilute in Isoton (a balanced salt solution from Curtis Matheson Science Inc., 357 Hamburg

Turnpike, Wayne, NJ 07470, USA) in a counting vial. Typically, we dilute 0.5 ml of the harvest from a T75 into 19.5 ml of Isoton. This sample is then counted in a Coulter counter. Other particle counting devices may also be used, as can a haemocytometer. After determining the number of cells in the harvest, add 40 ml of complete medium to each of the T75 flasks that are to be seeded with cells. Other cell culture vessels may also be used. Add 0.53 ml of complete medium per cm^2 of cell growth surface. Resuspend the cell harvest and inoculate the fresh culture vessels with the appropriate volume of cell harvest. We seed cultures at a constant density of 1×10^4 cells per cm^2 of growth surface.

On some occasions it is necessary to seed a large number of flasks or multiwell plates. To do this we prepare a suspension of cells and seed directly from this into the flasks or wells. Calculate the total volume of cell suspension required, and then double it. For example, if we are preparing to set up two 24-well plates where each well is 2 cm^2 in surface area (2×10^4 cells per well total) and will receive 1 ml of cell suspension, then we need to seed 48 wells. This requires 48 ml $\times$ 2×10^4 cells/ml $= 9.6 \times 10^5$ cells. However, if we were to try and seed 1 ml into each of 48 wells there would be problems with ensuring that there was a truly equal seeding of cells into all wells. To overcome this problem we routinely prepare twice as much of the cell suspension as is actually required (in this case it would be 96 ml with 1.92×10^6 cells). We also count the number of cells in an aliquot before we begin to seed the cells and then another aliquot when the seeding is completed. During the seeding it is important to keep the cell suspension well mixed. This can be done in one of two ways. First, the cell suspension can be placed in a sterile beaker with a sterile stirring bar and the beaker placed on a stir plate in the laminar flow hood. One millilitre aliquots are then removed with a 1-ml pipette and the wells are inoculated. In the other technique we use a sterile Erlenmeyer flask to prepare the cell suspension and swirl the suspension after every six wells are seeded. This is a little less cumbersome. However, care must be taken to ensure that the cells remain evenly suspended. Counting the cells before and after the multiwell plates are inoculated will allow you to determine how successful you were.

Other than serum, dissolved CO_2 in equilibrium with HCO_3^-, represents the principal buffer system in the medium. CO_2 is volatile and the gas phase must be adjusted to the proper pCO_2 to maintain the pH of the medium at 7.4. To accomplish this sterilize a 95% air : 5% CO_2 mixture by passing it through a sterile, cotton-filled $CaCl_2$ drying tube. From the drying tube, deliver the gas through a sterile pipette into the gas phase in the cell culture flask. Flush directly over the medium surface for a few seconds. Close the flask tightly and incubate at 37°C.

2.2.1 *Calculating the population doubling level*

In order to maintain an accurate record of the *in vitro* age of the cell culture it is necessary to calculate the increase in the cumulative population doubling level since the last time the cells were passaged. This calculation is made directly from the number of cells counted in the cell harvest. For example, one week after seeding a T75 flask with the standard inoculum of 7.5×10^5 cells, the harvest contained 6.0×10^6 cells. The population doubling increase is calculated by the following formula: $N_H/N_I = 2^X$ (where N_H = cell harvest number, N_I = cell inoculum number, and X = the number of population doublings). Therefore:

$$6.0 \times 10^6 \text{ cells}/7.5 \times 10^5 \text{ cells} = 2X$$

or

$$\log_{10} (6 \times 10^6) - \log_{10} (7.5 \times 10^5) = X \log_{10} 2$$

Therefore $X = 3$.

Add the increase in the population doubling level to the previous population doubling level to arrive at the current cumulative population doubling level (CPDL). We write this CPDL on the newly seeded flask along with the seeding date. Thus the flask is labelled with the population doubling level at seeding.

Expect little or no increase in cell number within 24 h of subcultivation. There are two reasons for this. First, 20–50% of the cells do not survive subcultivation. Second, the first parasynchronous round of DNA synthesis does not begin until approximately 15 h and the first mitotic cells do not appear until about 30 h.

In the designation of CPDL, cell losses occurring at subcultivation are ignored. One can estimate this error by assuming that 0.5–1.0 doubling may not be counted for each subcultivation. In our routine procedure, there are, on the average, three doublings per subcultivation. Thus between 10 and 20 doublings of the surviving population are not counted over the full life history. For historical reasons and for comparison with the data of others, we do not correct for this.

2.2.2 *Defining the 'phase out' of a culture*

In the introduction we discussed the finite replicative capacity of normal human cells in culture. It is important when studying the cell biology of these cultures to have an accurate and reproducible way of determining the age of the cultures. There are numerous well-documented physiological changes which occur as cultures go through the *in vitro* aging process. With this in mind, we have established a routine which ensures that the cultures which we handle attain their maximum proliferative capacity. This is often referred to as 'phase out', or 100% lifespan completed. The cell stocks are subcultivated on a weekly schedule. When the cells do not form a confluent sheet after one week they are refed by replacing the spent medium with fresh complete medium and gassed with the 5% CO_2 mixture. Every week the cultures are checked. If they have formed a confluent layer then they are subcultivated as described above. However, if they are still not confluent the cultures are refed as described. After three weeks of refeeding if the cells are not confluent the flask is harvested anyway. If the population has not doubled during the three weeks of refeeding (i.e. for a T75 flask, $N_H < 1.5 \times 10^6$ cells), then the cell line is considered to be phased out. In our laboratory WI-38 cells 'phase out' on average at CPDL 67.

Primary cultures are prepared directly from tissues, and the first confluent monolayer is designated PDL 1. Although this designation is subject to major errors, it is virtually impossible to estimate cell doublings in the primary explant.

3. PROPAGATION OF CELLS IN SERUM-FREE MEDIUM

The serum-free serial subcultivation of various lines of normal human fibroblast-like cells has been reported. Unfortunately, it is necessary to either supplement the medium with a complex lipid mixture that is delivered in the form of liposomes (3) (which are unstable and must be prepared immediately before use) or the medium must be supplemented with very large quantities of bovine serum albumin (10) (this acts as a lipid carrier, but also contains many unknown and known mitogens such as insulin-like growth factor-I). In our laboratory we have focused on the short-term growth of WI-38 cells in a defined, serum-free, growth factor-supplemented medium (4, 5). This allows us to take advantage of the relative ease of carrying our stock cultures in an FBS-supplemented medium while permitting us to then switch over to a chemically defined serum-free system at any point in the lifespan of the cultures in order to study aging and specific growth factor action.

3.1 **Materials**

3.1.1 *Growth factors*

WI-38 cells are responsive to a fairly large number of different mitogens, but maximum stimulation (i.e. equivalent to 10% FBS-supplemented medium) requires specific combinations of growth factors. In the following sections we will describe our methods for growing WI-38 cells through at least ten population doublings at a rate and to an extent equivalent to 10% FBS-supplemented medium. Our original growth factor formulation consisted of the following: partially purified platelet-derived growth factor (PDGF) at 3 μg/ml, epidermal growth factor (EGF) at 100 ng/ml, insulin (INS) at 5 μg/ml, transferrin (TRS) at 5 μg/ml (all from Collaborative Research Inc., Bedford, MA 01730, USA), and dexamethasone (DEX) at 55 ng/ml (Sigma). Although some of these preparations (PDGF, INS, TRS) are only partially purified, this mixture provides an excellent starting point for growth studies. After describing this system we will elaborate on more highly purified and less complex systems in Section 3.3.

3.1.2 *Growth medium and trypsinizing medium*

We use a modified form of the basal medium MCDB-104 (11) (formula number 82-5006EA, Gibco, 3175 Stately Rd, Grand Island, NY) which differs from the original formulation as follows: sodium pantothenate substituted for calcium pantothenate, without $CaCl_2$, without Hepes buffer, and without Na_2HPO_4. This formulation allows us to use two different buffer systems and to work with varying concentrations of $CaCl_2$. We have this medium prepared as a powdered mixture in packets, each sufficient to make up 1 litre of liquid medium. To 700 ml of deionized distilled water add one packet of the powdered medium. Add the following additional components in the order listed: 1.0734 g of $Na_2HPO_4 \cdot 12H_2O$, 1.754 g of NaCl, 1 ml of a 1 M $CaCl_2$ solution, and 1.176 g of $NaHCO_3$. Bring this to 1 litre with deionized distilled water. Filter-sterilize the medium through a

0.22-μm pore filter into sterile glass bottles. If the medium has a purple tint, it indicates that it has become alkaline (pH above ~7.6). This should be corrected by bubbling a stream of 5% CO_2:95% air through the medium before filter sterilization. Do not attempt to freeze this medium because as it freezes the pH rises. Upon thawing the medium, salts will occasionally precipitate out of solution and may not be redissolved. Properly prepared medium should have a pH of 7.3–7.5. We prepare the medium in small batches and keep it refrigerated for up to 3 weeks. If it is impractical to buy the powdered medium it can be prepared from the original formulation or our modification by following the procedures described by McKeehan *et al.* (11).

A variation of MCDB-104 must be made up to be used for the concentrated growth factor stock solutions. To 900 ml of deionized distilled water add one packet of the MCDB-104, 1.0743 g of $Na_2HPO_4 \cdot 12H_2O$, 11.9 g of Hepes buffer, 1 ml of 1 M $CaCl_2$, and 25 ml of 1 M NaOH. Adjust the pH of the solution to 7.5 by titration with 1 M NaOH and bring the volume to 1 litre with deionized distilled water. Sterilize by filtration through a 0.22 μm filter. This medium can be stored frozen until needed. This is the original formulation of MCDB-104 (11). Although WI-38 cells seem to tolerate the 50 mM Hepes it causes the formation of intracellular vacuoles (12), the consequences of which are unknown. Other cell lines may be even more sensitive to it. For these reasons we have adopted a bicarbonate-based medium for cell growth and only use the Hepes-based medium to make up 100× growth factor stock solutions.

Make up the growth factor stock solutions in the Hepes-buffered form of the medium, using sterile plastic pipettes, and store the peptide growth factors in sterile plastic test tubes. DEX should be stored in sterile siliconized glass test tubes. EGF at 2.5 μg/ml (100×), INS at 500 μg/ml (100×), TRS at 500 μg/ml (100×), and 5 mg/ml DEX in 95% ethanol then diluted to 5.5 μg/ml with the MCDB-104 (100×). PDGF must be handled differently because it readily adsorbs to glass surfaces. Follow the individual supplier's instructions and use PDGF at approximately 6 ng/ml if it is highly purified and at about 3 μg if it is only a crude, partially purified preparation. In our laboratory both the human and porcine forms of PDGF are equally potent as mitogens. All these stock solutions should be dispensed into 0.5–1.0 ml volumes and stored at -20°C for short periods (up to 4 weeks) or at -70°C for longer periods (3–4 months). Finally prepare a filter-sterilized solution of 1 mg/ml soybean trypsin inhibitor Type I-S (1×) (Sigma). We dispense this into 8-ml aliquots into sterile plastic tubes. Avoid repeated freeze–thaw cycles for all stock solutions.

Make up the trypsinizing medium exactly as described in Section 2.1.1. As mentioned before this should be kept on ice until you are ready to use it.

Prepare the exact amount of serum-free, growth factor-supplemented medium that is required for the number of tissue culture vessels that are to be seeded with cells. This medium must be prepared in a sterile plastic container such as a 50-ml capped centrifuge tube, or a tissue culture flask. Using a sterile plastic pipette or sterile plastic graduated cylinder dispense the appropriate volume of medium into the tissue culture vessels at 0.53 ml/cm^2. Close the caps on the vessels and lay the vessels on the floor of the laminar flow hood with cell growth surface down.

3.2 Trypsinizing and harvesting the cells

The cells are released from the growth surface exactly as described in Section 2.2. However, once the cells are released the procedure changes. Add 8 ml of the soybean trypsin inhibitor solution to the flask (this is instead of 8 ml of complete MEM, FBS-containing medium), and pipette gently up and down in order to form a single cell suspension. Remove a 0.5-ml aliquot of the cell suspension and count the number of cells as described in Section 2.2. Using this information calculate the CPDL of the culture as described in Section 2.2.1.

The cells must now be removed from the soybean trypsin inhibitor solution. Wash the cell harvest once by centrifugation at 75 g for 5 min at 4°C. Resuspend the cells in 7–10 ml of serum-free growth factor-free $NaHCO_3$-buffered MCDB-104. After the centrifugation you will find that you have only recovered 50–70% of the original cell number. Higher speeds or longer centrifugation times may recover more cells but there will be a decrease in viability. Therefore we recommend the short low-speed spin described above. Remove a 0.5-ml aliquot from the resuspended cell harvest and count the number of cells. This will allow for the proper seeding density to be achieved. With experience you will be able to estimate the cell loss during centrifugation and then be able to judge the appropriate volume in which to resuspend the cells so that: (i) the cells will be concentrated enough to give reliable counts on the Coulter counter; and (ii) the inoculum volume will be small compared to the volume of growth factor supplemented medium in the flasks and thus have only a minimal dilution effect (5% or less) on the concentrations of growth factors. Then, using a plastic pipette seed the flasks as described in Section 2.2. Follow this by gassing the flasks with the 5% CO_2:95% air mixture, close the caps tightly and place in the CO_2 incubator (Section 2.2).

For studies that involve growth curves it is both convenient and economical to grow the cells in multiwell plates (e.g. 24-well plates). Essentially the same procedure is used in this case. The wells are approximately 2 cm^2 and receive 1 ml of growth factor-supplemented medium. When the cells are resuspended after the centrifugation they must be resuspended such that there are 2×10^4 cells per 25–50 μl. Then, using a variable micropipette (e.g. Eppendorf or Rannin) which has been wiped down with 70% ethanol, and a sterile tip, inoculate the wells with the appropriate number of cells. Place the multiwell plates in the 37°C CO_2 incubator.

3.2.1 *Alternate growth factor formulations*

In the preceding section we described our original growth factor formulation for culturing WI-38 cells under serum-free conditions. The combination of PDGF, EGF, INS, TRS and DEX supports growth at a similar rate and to a final cell density as does 10% FBS-supplemented medium. A major drawback to this formulation is the expense of even partially purified PDGF. The major effect that PDGF has on these cells, as well as other normal human fibroblasts, is to drive them on to a higher saturation density. PDGF can be left out of the medium with very little effect upon the growth rate for the first few days of culture. Without PDGF the final saturation density is about 70% (7×10^4 cells/cm^2) of that obtained with it (1×10^5 cells/cm^2). Since EGF, INS, TRS and DEX support low density cell growth as well as FBS, there is an added advantage to leaving PDGF out of the formulation. There are fewer mitogenic components and therefore it is a less complex system to study.

INS is an effective mitogen for many cell types, and it is now known that INS's mitogenic action is derived from its ability to bind, at high concentration (and low affinity), to the insulin-like growth factor-I (IGF-I) receptor (13). The IGF-I receptor actually mediates the mitogenic response to both of these growth factors. IGF-I is maximally effective with WI-38 cells at 100 ng/ml, that is at a 50-fold lower concentration than INS. IGF-I can be freely substituted for INS in the growth factor mix, however, it is also very expensive and comparable growth can be obtained using INS (14).

Finally, it is possible to eliminate the TRS from the serum-free medium. To do this an FeSO$_4$ stock solution must also be prepared. Make up a 1-litre solution of 1 mM FeSO$_4$ (200×) and add 1 drop of concentrated HCl. Filter-sterilize this through a 0.22 μm filter. This solution can be stored at room temperature for approximately 2 months. Discard the solution at the first hint of colour change (11). This solution is used to supplement the basal medium just before the growth factors are added. In our original formulation we used TRS, the plasma iron transport protein, as the vehicle for getting iron into the cells. Subsequently we and others (5, 15) observed that if slightly acidic FeSO$_4$ is added to the medium then TRS is not required. This apparently is dependent on the oxidation state of the iron. In the reduced state it apparently passes freely into the cell, but when it is oxidized (which occurs with time) it requires the transport protein TRS in order to enter the cell.

3.2.2 Special precautions for serum-free cell growth

Although serum-containing medium contributes a large number of unknown variables to the tissue culture system it also provides protection for fastidious cells. For example, serum can adsorb heavy metals or other contaminants which may be found in the deionized water used to prepare the basal medium. Therefore, water quality becomes critical when cells are switched over to a serum-free system such as the one described above. We use deionized, glass-distilled water for our serum-free experiments and find it satisfactory.

It is important that the cells be harvested using the soybean trypsin inhibitor rather that FBS-containing medium (FBS contains trypsin inhibitors). This is because mitogens present in the serum may be sequestered inside the cells or associated with their plasma membranes. The presence of even small quantities of serum complicates the interpretation of cell growth response data.

The other important consideration when working with serum-free growth factor-supplemented medium is ensuring that the growth factors are not lost by adsorption to glass surfaces. We use only sterile plastic, centrifuge tubes, pipettes, flasks, etc. when handling peptide growth factors. Some, such as EGF are

Table 3. A classification of growth factors which stimulate WI-38 cell proliferation.

Class I		Class II		Class III	
EGF	(25 ng/ml)	IGF-I	(100 ng/ml)	HC	(55 ng/ml)
FGF	(200 ng/ml)	IGF-II	(500 ng/ml)	HC	(55 ng/ml)
PDGF	(6 ng/ml)	INS	(5 μg/ml)		
THR	(500 ng/ml)				

relatively easy to work with, while others such as IGF-I and PDGF are extremely 'sticky' and must be handled only with plastic pipettes and tubes and with the minimum of manipulations.

3.3 A classification of growth factors for WI-38 cells

There are a number of growth factors which stimulate WI-38 cell proliferation, and we have found that they can be functionally grouped into three classes (16). These are shown in *Table 3*. Class I includes EGF, fibroblast growth factor (FGF), PDGF, and thrombin (THR). Class II includes IGF-I, IGF-II (or the rat homologue multiplication stimulating activity, MSA) and INS. Class III includes hydrocortisone or its synthetic analogue DEX. At low cell density, members of each of the three classes act synergistically in stimulating cell proliferation. Any class I mitogen in combination with any class II and either class III mitogen stimulate growth to an extent similar to FBS-supplemented medium. By using this scheme it should be possible to employ a variety of growth factors in order to maximally stimulate various types of human fibroblast cultures derived from a variety of anatomical sites.

4. CONCLUSIONS

We have presented the methods which we have developed over some 20 years for the reproducible and optimum growth of human diploid fibroblasts in both serum-containing and serum-free media. The major advantage to serum-containing medium is the ease with which the cells may be subcultivated in order to achieve their maximum proliferative potential. The advantage of the serum-free medium is its completely defined composition and the accompanying ability to rigorously control the regulation of cell proliferation, and presumably other hormonally regulated functions.

5. REFERENCES

1. Jensen, F., Koprowski, H. and Ponten, J. A. (1968) *Proc. Natl. Acad. Sci. USA*, **50**, 343.
2. Zimmerman, R. J. and Little, J. B. (1983) *Cancer Res.*, **43**, 2183.
3. Bettger, W. J., Boyce, S. T., Walthall, B. J. and Ham, R. G. (1981) *Proc. Natl. Acad. Sci. USA*, **78**, 5588.
4. Phillips, P. D. and Cristofalo, V. J. (1980) *J. Tiss. Culture Meth.*, **6**, 123.
5. Phillips, P. D. and Cristofalo, V. J. (1981) *Exp. Cell Res.*, **134**, 297.
6. Hayflick, L. and Moorhead, P. S. (1961) *Exp. Cell Res.*, **25**, 585.

7. Hayflick, L. (1965) *Exp. Cell Res.,* **37**, 614.
8. Cristofalo, V. J. and Sharf, B. B. (1973) *Exp. Cell Res.,* **76**, 419.
9. Cristofalo, V. J. and Charpentier, R. (1980) *J. Tiss. Culture Meth.,* **6**, 117.
10. Yamane, I., Kan, M., Hoshi, H. and Minamoto, Y. (1981) *Exp. Cell Res.,* **134**, 470.
11. McKeehan, W. L., McKeehan, K. A., Hammond, S. L. and Ham, R. G. (1977) *In Vitro,* **13**, 470.
12. Verdery, R. B., Nist, C., Fujimoto, W. Y., Wight, T. N. and Glosmet, J. A. (1981) *In Vitro,* **17**, 956.
13. Van Wyk, J. J., Graves, D. C., Casella, S. J. and Jacobs, S. (1985) *J. Clin. Endocrinol. Metab.,* **61**, 639.
14. Phillips, P. D., Pignolo, R. J. and Cristofalo, V. J. (1988) *J. Cell. Physiol.,* **133**, 135.
15. Walthall, B. J. and Ham, R. G. (1981) *Exp. Cell Res.,* **134**, 303.
16. Phillips, P. D. and Cristofalo, V. J. (1988) *Exp. Cell Res.,* **137**, 396.

CHAPTER 9

Cell synchronization

GARY S. STEIN and JANET L. STEIN

1. INTRODUCTION

There is a growing recognition that our understanding of cell growth control necessitates the identification and characterization of cellular, biochemical and molecular events that occur during specific stages of the cell cycle. Thus, synchronized cells provide the essential model systems for examining the regulation of cell proliferation. In this chapter we will present methods that have proven to be effective for synchronizing continuously dividing cells and methods for monitoring cell synchrony.

2. SYNCHRONIZATION OF CONTINUOUSLY DIVIDING CELLS

2.1 Double thymidine block

Synchronization of continuously dividing cells can be achieved by imposing a metabolic block that meets the following criteria: (i) cells are arrested only at one specific stage of the cell cycle; that is, progression through the cell cycle continues until the stage where the block occurs; and (ii) the block is reversible for permitting resumption of cell cycle traverse with minimal perturbations of biochemical, cellular or molecular parameters of proliferation.

Historically, the use of excess thymidine was the first widely accepted, and remains one of the most effective, method for inducing cell synchrony (1). By treatment with two sequential 'thymidine blocks' a synchronous population of cells can be obtained at the beginning of S-phase, and the method can be effectively utilized to synchronize both suspension and monolayer cells. It should be emphasized that for this method to function optimally all cells in a population to be synchronized must be undergoing exponential growth. A description of the double thymidine block procedure has been reported by Stein and Borun in 1972 (2).

The protocol described below for synchronization of exponentially growing HeLa S3 cells in suspension cultures is based on a doubling time of 20 h with the following times associated with each stage of the cell cycle: G_1, 6 h; S-phase, 9 h; G_2, 4 h; and mitosis, 1 h.

(i) Cells at a concentration of 5×10^5/ml are diluted with fresh medium to a final concentration of 3.5×10^5 cells/ml, and thymidine is added to a final concentration of 2 mM from a 50× stock solution prepared in 'Spinner Salts' [Spinner Salt

Solution (Eagle), Gibco Laboratories, Grand Island, NY]. The thymidine stock solution can be sterilized by autoclaving.

Rationale: a 12–16 h treatment of exponentially growing HeLa cells with 2 mM thymidine will allow all cells that were in G_2, mitosis or G_1 at the initiation of the block to progress to the G_1/S-phase boundary, while cells undergoing DNA replication will be immediately arrested in S-phase. Thus, at the completion of the first thymidine block, approximately 50% of the cells will accumulate at the G_1/S-phase transition point and the other half of the cells will be arrested at various points in S-phase.

(ii) Release of the first thymidine block is accomplished by pelleting the cells by centrifuging at 600 *g* for 5 min, carefully pouring off the growth medium to maximally eliminate thymidine-containing medium in contact with the cell pellet, washing the cells in 200 volumes of 'Spinner Salts' at 37°C, and resuspending the cells in fresh growth medium at a final concentration of 3.5×10^5 cells/ml. Care should be taken to maintain the cells at 37°C throughout the synchronization procedure, because even a slight decrease in temperature for several minutes will result in significant delays in cell cycle progression. Optimal conditions for cell synchronization are achieved by carrying out the entire procedure in a 37°C environmental room using an ambient temperature centrifuge for harvesting the cells.

Rationale: a 9-h release period permits all cells to exit S-phase. The 50% of the cells that accumulated at the G_1/S-phase boundary during the first thymidine block now will be in the initial segment of G_2, and those cells that were blocked at the various points during S-phase will be distributed between G_2, mitosis and G_1.

(iii) After completion of the 9-h release period, a second thymidine block is initiated by diluting the suspension cultures with fresh medium to a final concentration of 3.5×10^5 cells/ml and adding thymidine to a final concentration of 2 mM.

Rationale: a 12–16 h treatment with excess thymidine accumulates all cells at the G_1/S-phase boundary.

This double thymidine block procedure can be carried out equally effectively using monolayer cultures. However, it is essential that the cell density at the time of the initial block be such that active growth can be maintained throughout the time course of the synchronization procedure. Blocks are initiated by pouring off the growth medium and providing fresh medium with 2 mM thymidine. Release of the thymidine blocks involves pouring off the thymidine-containing medium and washing the monolayers with an equal volume of 'Spinner salts' (at 37°C) prior to feeding with normal growth medium.

2.1.1 *Use of other S-phase inhibitors*

Several variations of the double metabolic block procedure have been developed that are effective for synchronization of continously dividing cells. Aphidicolon at a concentration of 5 μg/ml (3) or hydroxyurea at a final concentration of 1 mM can be substituted for the second thymidine block.

2.2 Mitotic selective detachment

This procedure yields mitotic cells that will synchronously progress through G_1 and then enter S-phase. During mitosis, monolayer cells assume a more rounded shape which results in considerably less of the cell surface being in contact with the culture vessel. The mitotic cells are readily detached from either glass or plastic surfaces (4). Although the yield of mitotic cells is rather low with this procedure, early G_1 events can be examined without the complications that arise due to the decay of synchrony that occurs by the time the S-phase cells synchronized by double thymidine block have passed through S-phase, G_2 and mitosis. Additionally, the mitotic selective detachment procedure permits a population of synchronized cells to be obtained without the potential problems associated with chemical inhibitors.

The following procedure has been developed for obtaining selectively detached mitotic HeLa S3 cells, and it can be readily adapted for other cell lines. The protocol presented is based on initiating the synchronization procedure with 4 litres of exponentially growing suspension cells at a concentration of 5×10^5 cells/ml in Joklik-modified Eagle's minimal essential medium supplemented with 7% calf serum. All procedures should be carried out in a 37°C environmental room.

(i) Cells are pelleted by centrifugation at 600 g for 5 min and resuspended in 2 litres of Eagle's minimal essential medium containing 7% calf serum. 50-ml aliquots of the cells are added to each of 40 1-litre borosilicate glass Blake culture flasks. The cells are gassed with a gentle stream of 95% air:5% CO_2 through a cotton-plugged, sterile pipette that is inserted through the open neck of the culture vessel. Then, the Blake bottles are closed with a silicon or neoprene rubber stopper and placed horizontally on a perfectly level surface to permit even plating of the cells.
Rationale: by transferring the HeLa cells from Joklik-modified Eagle's minimal essential suspension medium, which is calcium-free, to Eagle's minimal essential medium, which contains calcium and a reduced level of phosphate, adherence of the cells to the surface of the culture vessels is promoted.

(ii) Six hours after plating the cells, the culture flasks are shaken vigorously to remove loosely attached cells. The medium is poured off and discarded, and the monolayers are rinsed twice with 15 ml of serum-free Joklik-modified minimal essential medium. 50 ml of Joklik-modified minimal essential medium containing 7% calf serum is then added to each culture, and the vessels are gassed with 95% air:5% CO_2.
Rationale: replacement of calcium-free 'suspension medium' for calcium-containing medium will facilitate detachment of the cells from the surface of the culture vessel when they enter mitosis and 'round up'.

(iii) Nine hours after plating, the culture vessels are each picked up and gently agitated to facilitate detachment of the mitotic cells, which are pooled and harvested by centrifugation at 600 *g* for 5 min. The mitotic cells are resuspended at 4×10^5 cells/ml in Joklik-modified minimal essential medium supplemented with 7% calf serum and maintained in suspension culture. The yield of mitotic cells can be readily determined by counting in a haemocytometer, and the mitotic figures are evident under phase-contrast microscopy.

2.2.1 *Enhanced yield of selectively detached mitotic cells*

Two approaches have proven to increase significantly the yield of mitotic cells by the selective detachment method. The first is to carry out a single thymidine block prior to mitotic detachment. This protocol involves treating suspension cells in Joklik-modified minimal essential medium with thymidine at a concentration of 2 mM for 12 h prior to transfer to calcium-containing Eagle's minimal essential medium and plating in Blake culture flasks. The second approach is to add colcemid to the monolayers 7 h after plating and collecting the cells 6 h later.

The thymidine and colcemid effects are generally considered to be reversible. However, this cannot be unequivocally assumed and appropriate controls should be carried out for all parameters to be assayed.

3. MONITORING CELL SYNCHRONY

Two of the most straightforward approaches to monitoring cell synchrony are to determine, at frequent intervals, the rate of DNA synthesis by measuring [^{3}H]thymidine incorporation into trichloroacetic acid (TCA)-precipitable material or to establish the percentage of the cell population undergoing DNA synthesis by [^{3}H]thymidine incorporation followed by autoradiography. Although time-consuming, autoradiography provides the most direct approach to following cell synchronization, since visual examination of the autoradiographic preparations unquestionably indicates which cells are replicating DNA. The specific protocols for autoradiography will not be presented here since they have been described in an indepth and comprehensive manner by Baserga and Malamud (5). What follows is a rapid method for measuring the rate of DNA synthesis in suspension and monolayer cultures.

3.1 **Rate of DNA synthesis in suspension cultures**

   (i) [^{3}H]thymidine is added to 2 ml of cells to a final concentration of 5 μCi/ml, and the cells are incubated at 37°C for 30 min with gentle agitation.
  (ii) Cells are pelleted by centrifugation at 600 *g* for 5 min and the medium is removed.
 (iii) The cell pellet is washed twice in ice-cold 'Spinner Salts' followed each time by centrifugation at 4°C at 600 *g* for 5 min.
  (iv) The cell pellet is resuspended in 6 ml of cold 10% TCA and maintained in an ice bath for at least 5 min.
   (v) The precipitate is collected on 2.5 cm Millipore type HA filters that have

been presoaked in 10% TCA. This is readily carried out by placing the presoaked filters in a vacuum filtration manifold, pouring the precipitate into the manifold well and allowing the TCA to pass through the filter. The filters are then washed three times with 5 ml of 10% TCA.

(vi) After the last TCA wash has passed through the filters, the filters are removed and placed at the bottom of a glass liquid scintillation counting vial.

(vii) The filter is solubilized by addition of 1 ml of Cellusolve (ethylene glycol monoethyl ether).

(viii) 10 ml of a Cellusolve liquid scintillation counting cocktail is added to each vial containing the solubilized filter [toluene : Cellusolve : Liquifluor (72 : 24 : 1, by vol.)].

3.2 Rate of DNA synthesis in monolayer cultures

(i) [^{3}H]Thymidine is added to the culture medium to a final concentration of 5 μCi/ml and the cells are incubated at 37°C for 30 min.

(ii) The culture medium is removed by aspiration and the monolayer is rinsed twice with ice-cold 'Spinner Salts'.

(iii) The cells are scraped from the culture vessel in 'Spinner Salts' and transferred to 7-ml centrifuge tubes.

(iv) The cells are pelleted by centrifugation at 4°C at 800 g for 5 min and the medium is discarded.

(v) TCA precipitation, collection of precipitates on Millipore filters and assessment of incorporated radioactivity are carried out as described above for determining rate of DNA synthesis in suspension cells.

4. REFERENCES

1. Bootsma, D., Budke, L. and Vos, O. (1964) *Exp. Cell Res.*, **33**, 301.
2. Stein, G. S. and Borun, T. W. (1972) *J. Cell Biol.*, **52**, 292.
3. Pedrali-Noy, G., Spadari, S., Miller-Faures, A., Miller, A. O. A., Kruppa, J. and Koch, G. (1980) *Nucleic Acids Res.*, **8**, 377.
4. Terasima, T. and Tolmach, L. J. (1963) *Exp. Cell Res.*, **30**, 344.
5. Baserga, R. and Malamud, D. (1969) *Autoradiography*, Hoeber, New York.

CHAPTER 10

Growth and maintenance
of BALB/c-3T3 cells

WALKER WHARTON and MIRIAM J. SMYTH

1. INTRODUCTION

BALB/c-3T3 cells are a non-transformed embryonic murine mesenchymal line isolated and first characterized by Todaro and Green (1). One of the subclones (A31) of the original cell line has been widely utilized in experiments in several areas of cellular and molecular biology. Even though both immortal and aneuploid, BALB/c-3T3 cells can be maintained in a state in which they exhibit stringent density-dependent growth inhibition and very rigorous growth factor requirements. It is straightforward, however, to obtain variant subclones with various degrees of neoplastic transformation. These cells arise either spontaneously or following treatment with any one of a number of agents ranging from chemicals to viruses. These attributes make BALB/c-3T3 cells particularly valuable in studies of growth regulation and transformation at a cellular, biochemical and molecular level. This chapter, which consists of procedures necessary to grow and utilize BALB/c-3T3 cells, is divided into two major sections. The first describes the basic culture techniques necessary to maintain stock cultures. The second section describes more specialized techniques by which these cells are used to measure aspects of growth and transformation.

2. BASIC TISSUE CULTURE TECHNIQUES

2.1 Solutions used in the culture of BALB/c-3T3 cells

3T3 cells are typically maintained in a combination of a culture medium (a defined mixture of salts, amino acids, vitamins, and glucose) and serum (a melange of mitogens, hormones, maintenance factors and non-specific proteins). Details of these components are provided below.

2.1.1 Medium

Two basic variations of the original Eagle's minimal essential medium (MEM) can be used to grow 3T3 cells; alpha-modified MEM (α-MEM) or Dulbecco's modified MEM (DMEM). Both of these are considerably richer than MEM, since they include a number of vitamins as well as both essential and non-essential amino acids. Either of these two media proves satisfactory, since in side-by-side growth curves we have found identical doubling times and saturation densities. Less complex media, such as F-10, are not suitable for supporting the growth of 3T3 cells. Medium can be purchased either in single strength or concentrated liquid form, or as a powder. The water used to dilute concentrates or to dissolve powders should be both distilled and deionized, and there should be some form of monitoring system in place to ensure proper water quality.

Glutamine is essential for growth and is included in the formulation of both αMEM and DMEM. However, this particular amino acid is not very stable. To ensure effective concentrations, a new aliquot should be added to fresh medium as it is being used to feed cells. Glutamine is made up as a 30 mg/ml stock in distilled water, sterile filtered, and stored at −20°C. It should then be added at 10 ml/litre of 1× medium.

Antibiotics are sometimes added to the culture medium to help prevent inadvertant bacterial contaminations. The most popular choice is a combination of penicillin (50 U/ml) and streptomycin (50 μg/ml), although gentamicin (50 μg/ml can also be added either in combination or alone. We have not observed a growth inhibitory effect of these antibiotics on 3T3 cells, although this can sometimes be a problem with other cell types. BALB/c-3T3 cells are, however, very sensitive to the toxic effects of antifungal agents, so the prophylactic use of drugs such as nystatin or amphotericin should be avoided. If cultures become accidently infected with a fungus, every attempt should be made to obtain duplicate uncontaminated cells before attempting to eradicate the fungus with drugs.

Media can be buffered in a number of ways, but a combination of bicarbonate and CO_2 is the most satisfactory for 3T3 cells. The quantity of bicarbonate added to media should be such as to maintain a pH of 7.5 in an atmosphere containing 5% CO_2. Instructions providing amounts of bicarbonate appropriate for the specific medium are supplied with the packing literature. Hepes-buffered medium, a popular alternative with some cells, causes a significant inhibition in the growth rate of BALB/c-3T3 cells.

2.1.2 *Serum*

Serum, typically added to a defined medium at a final concentration of 10%, provides a rather ill-defined supply of mitogens and attachment factors. Particular batches of both calf and foetal calf serum (FCS) will maintain excellent growth rates and, in some instances, a 50:50 mixture of the two sera can be used as an economy measure. For optimal results, observe the growth characteristics with small samples of several batches of serum and purchase a large quantity of the particular batch that gives the best results. Calf, but not foetal calf, serum should be heat inactivated (56°C, 30 min) prior to use.

In order to completely or partially replace serum, defined hormonal supplements can either be designed or obtained commercially. Some of the commercial materials are advertised as being adequate to support the growth of BALB/c-3T3 cells at a rate equivalent to that seen in medium containing 10% serum. However, they often contain levels of components such as epidermal growth factor (EGF)

Table 1. Preparation of trypsin–EDTA solution.

Compound	g/litre concentration
NaCl	8.0
KCl	0.4
Glucose	1.0
NaHCO$_3$	0.35
Trypsin	1.25
EDTA, tetrasodium, dihydrate	0.2
Phenol Red (5 mg/ml)	0.5 ml

1. Mix all the components together and stir for approximately 30 min.
2. Not all the trypsin will dissolve.
3. Adjust the pH to 7.2 and sterile-filter.
4. Store at −20°C.

that cause significant receptor down-regulation. This could be an important variable when preparing cultures for some types of experiments in which ligand binding is measured. It is obvious that a completely defined medium would be a benefit in reducing experimental variability, but with present technology it is not clear that it is worth the extra expense, especially for the routine support of stock cultures.

2.1.3 *Trypsin–EDTA*

3T3 cells are detached from plastic culture ware by treatment with a Puck's Saline A solution containing both trypsin and EDTA. The instructions for making up such a solution are given in *Table 1*.

2.2 **Growth of BALB/c-3T3 cells**

2.2.1 *Stock cultures*

Stock cultures of BALB/c-3T3 cells are maintained in 100-mm tissue culture dishes. These stock plates serve as a source of both new stocks as well as cells to be plated into cultures used in experiments. We have found empirically that rates of spontaneous transformation are reduced markedly if the initial plating density is reduced from that originally suggested by Todaro and Green (1). The 3-day transfer schedule they described should, however, be rigorously enforced. Under this regime, cells can be continously maintained for up to a 3-month period before a culture needs to be established.

(i) Cells should be seeded into ten plates containing medium supplemented with 10% FCS at an initial density of 5 × 10^4 cells per dish.
(ii) Following a 3-day period the medium of eight of the initial plates should be removed and replaced with fresh medium containing FCS. These plates will

eventually be allowed to reach confluence and will be used to establish experimental cultures.

(iii) To regenerate a supply of ten new stock plates, remove the medium from the other two plates and add 5 ml of trypsin–EDTA. Following a 30-sec incubation at room temperature, aspirate all but about 0.25 ml of the trypsin solution and place the plates in a 37°C incubator for approximately 5 min, or until observation using an inverted microscope indicates that the cells are completely detached. Add 5 ml of medium plus FCS to each plate, pipette up and down to disperse the cells evenly, and pool cells from the two plates. Remove an aliquot of the cell slurry, determine the cell concentration, and add an appropriate volume to each of ten new plates such that the final cell concentration is 5×10^4 cells per plate.

(iv) At the end of another 3-day incubation the medium is changed on the initial set of plates for the second time, on a set of eight 3-day-old plates for the first time, and the two remaining 3-day-old plates are used to set up ten new stock plates as described above.

(v) At the end of each 3-day incubation the entire process is repeated. After approximately 9 days of growth the cells on a plate reach confluence and become growth-arrested. They can be used to establish experimental cultures over about a 6-day period following confluence. Stock cultures, however, should not be established from cells that have ever been density-arrested. For reasons that are not yet clear, such a procedure will result in a high degree of spontaneous transformation.

2.2.2 *Seeding of cultures for use in experiments*

Stock plates that have reached confluence are utilized to seed cultures that will be used directly in experiments.

(i) Aspirate the medium from confluent 100-mm tissue culture dishes and add 5 ml of trypsin–EDTA. Immediately aspirate the trypsin solution and add another 5 ml of trypsin–EDTA. Following a 30 sec incubation at room temperature remove all but approximately 0.25 ml and place the plate at 37°C for about 5 min or until all the cells are detached. When using confluent cells, the extra trypsin wash will speed up the detachment and result in significantly fewer clumps.

(ii) Add 5 ml of medium plus 10% FCS to each plate, pipette up and down to disperse the cells, and pool the slurry from all the plates. Remove an aliquot and determine the cell number.

(iii) If experiments are designed to use density-arrested cells, cultures should initially be seeded at a density of 5×10^3 cells/cm^2. To maximize uniformity of plating conditions within the same experiment, one large volume of medium containing an appropriate cell concentration should be prepared and then divided into individual plates. This suspension is gently swirled before each aliquot is withdrawn. For standard tissue culture plates, use the information in *Table 2* to calculate the volume and number of cells needed to establish the necessary plates for any given experiment.

Table 2. Conditions for plating cells for experiments.

Plate diameter (mm)	Area (cm²)	Volume of medium (ml)	Cells/plate (× 10⁻⁴)
35	9.6	2	4.8
60	28.3	5	14.1
100	78.5	10	39.2
150	176.6	30	88.3

(iv) The medium should be changed 3 days following plating, and density-dependent growth arrest will be achieved following a further 4–5 days of culture. Depending on the specifics of the experiment, the cells can be used over the next 2- to 3-day period.

(v) Experiments investigating the effects of compounds or conditions on the logarithmic growth rate of 3T3 cells can be performed efficiently in 60-mm tissue culture plates. Cells should be seeded at an initial density of 2×10^4 cells per plate. After a 24-h period, allowing for attachment, the medium can be changed to expose the cells to the relevant experimental conditions. Logarithmic growth will be maintained in control cultures for at least 6 days.

2.3 Permanent storage of cells

2.3.1 Cryopreservation

Cells can be indefinitely frozen in a vapour phase over liquid nitrogen at −130°C and for at least several months in a −70°C freezer with no loss of viability or alteration in phenotype. The steps critical in this process are designed to prevent damage by ice crystals and primarily involve the elimination of clumps, suspension of cells in proper solutions, and lowering of the temperature at a suitable rate. Recovery following freezing is highest when the cultures to be harvested are in a logarithmic growth phase (we use plates that are about 60% confluent) and have had fresh medium added within a 24-h period. The following procedure uses volumes suitable for cells grown on 100-mm dishes.

(i) Remove cells from plates by treatment with trypsin–EDTA as described in Section 2.2.1.

(ii) Resuspend the cells in 5 ml of MEM containing 10% FCS, pipette the suspension up and down to break up clumps, and pellet the cells with a low speed centrifugation at room temperature.

(iii) Resuspend the cell pellet in MEM plus 20% FCS, using 0.75 ml of medium for each p100 originally harvested.

(iv) Add an equal volume of MEM containing 20% FCS and 15% DMSO to the cell slurry and gently mix thoroughly. This freezing solution can be made up in 100-ml batches, sterilized by filtration, and frozen in 10-ml aliquots.

(v) Slowly cool the cells to −70°C. This can be accomplished in many ways, including the use of very sophisticated and expensive instruments that can be programmed to lower the temperature according to a predetermined pattern. We have had the best results with a very simple protocol. Place the vials in the styrofoam container supplied with 15-ml Corning conical tubes. Invert a second styrofoam holder and place it over the first, such that the vials are completely enclosed. Seal the holders in a thick plastic bag and place at −70°C for 48 h. The vials can then be placed directly in a liquid nitrogen freezer for long-term storage.

(vi) To ensure the quality of each freeze down, one vial should be thawed and cultured after it has been in the nitrogen freezer for 48 h.

2.3.2 Thawing cells

In contrast to freezing, in which the temperature should be lowered slowly and steadily, frozen vials of cells should be thawed as quickly as possible and diluted into fresh medium to prevent the toxic effects of dimethylsulphoxide (DMSO).

(i) Place a frozen vial of cells in a 37°C water bath and agitate steadily until the entire sample is thawed.

(ii) Add the 1.5-ml sample to 10-ml of warm medium containing 10% FCS and centrifuge for 5 min at 1000 r.p.m. in a clinical centrifuge.

(iii) Resuspend the pellet in fresh complete medium and add to a flask containing prewarmed medium. Cells should start to adhere within 2 h. In a good-quality freeze-down, over 90% of the cells should be attached to the bottom of the flask after 24 h.

2.4 Mycoplasma

Pleuropneumonial-like organisms (PPLO) can be a particularly bothersome problem since infection can cause artifacts and significant problems in experimental stability. Regular checks of stock cultures should be made, and all cells received from outside laboratories should be screened before being extensively used.

Some antibiotics can be effective in lowering mycoplasma contamination, but a complete cure is rare. Unless the infected cells are quite valuable and unique it is always advisable to dispose of contaminated cultures and obtain new cells.

There are several methods for detecting potential problems with PPLO, but the following is straightforward and has the advantage of actually culturing the organism. It is a variation of the procedure proposed by House and Waddell (2).

(i) Suspend 250 g of active bakers yeast in 1 litre of distilled water, adjust the pH to 4.5, and heat the suspension to 80°C. Centrifuge at 3000 r.p.m. and remove the supernatant. Adjust the pH to 7.5, sterile-filter, aliquot into 10-ml screw-cap tubes, and store at −70°C.

(ii) Prepare Difco Bacto PPLO agar according to the directions on the bottle. Divide it into 70-ml quantities, sterilize by autoclaving and store at 5°C.

(iii) Divide sterile, non-activated horse serum into 20-ml aliquots and store at −20°C.

(iv) Melt the 70 ml of agar in a boiling water bath, and place in a 45°C water bath for at least 30 min. Add 10 ml of yeast extract and 20 ml of horse serum that have been pre-equilibrated at 37°C and quickly pipette 5 ml into 60-mm Petri dishes before the temperature falls below 45°C. Allow the plates to dry with the lids ajar for about 1 h.

(v) Obtain a sample of tissue culture medium containing some cells from a plate that has not had a medium change for at least 3 days. A drop of this medium is placed in the centre of the agar plate with a sterile Pasteur pipette. When the medium is completely absorbed into the agar (~1 h) place the plates in a desiccator, with water in the bottom, and incubate at 37°C.

(vi) Look for signs of infection at 3, 5, and 7 days. The most likely location of a colony is at the edge where the inoculum meets the medium.

2.5 Spontaneous transformation

2.5.1 *Qualitative assays for spontaneous transformation*

Although BALB/c-3T3 cells are an immortalized line that will not senesce, the continuous use of a single culture will eventually be limited by the generation of spontaneous transformants. These cells arise by low frequency mutation-like events and exhibit at least some aspects of the transformed phenotype. As described earlier, we have empirically found that the generation of transformants can be limited by strict adherence to a low plating density in the stock cultures. There should, nevertheless, be a system on hand to monitor for the emergence of such cells.

One way to prevent interference by transformed cells is to have a strict regular schedule according to which new vials of cells are thawed. If cultures are discarded after 5–6 weeks, it is unlikely that there will be severe problems with the emergence of variants. However, this is not a maximally efficient use of frozen stocks and it is sometimes better to throw out cultures only when there is an indication of small numbers of transformants.

Transformed cells typically do not become density-arrested as readily as parent BALB/c-3T3 cells. A steady increase over time in the background mitogenesis or partial abrogations in growth factor requirements are indications that transformants are being generated.

Confluent transformed cells often have an elongated morphology, quite distinct from the epithelial-like non-transformed BALB/c-3T3 cells (*Figure 1*). Plates should be regularly scanned with an inverted microscope to determine if patches of cells with an altered morphology are appearing.

Once a culture becomes contaminated with spontaneous transformants, it is not useful for many of the experiments typically carried out with 3T3 cells. Although discarding the cultures and thawing new vials provide a short-term relief, it will eventually be necessary to reclone a culture to isolate homogeneous, non-transformed cells. The following protocol is based on the preparation of a solution in which the cells are diluted to such an extent that there is a negligible probability that an aliquot will contain two cells.

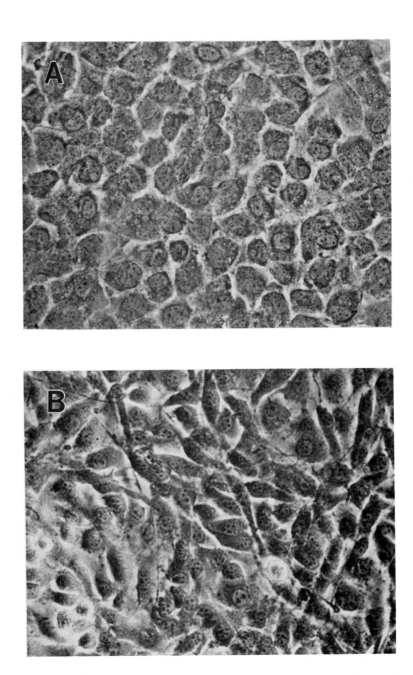

Figure 1. Characteristic morphology of parent BALB/c-3T3 cells (**A**) and a spontaneously transformed subclone (**B**) at a magnification of 320×.

2.5.2 *Procedure for recloning by limiting dilution*

(i) Remove the cells from a 10-mm culture plate by treatment with trypsin–EDTA as described in Section 2.2.1.

(ii) Resuspend the cells in 10 ml of αMEM containing FCS. Remove an aliquot and determine the cell concentration.

(iii) Dilute the cells into the medium with FCS such that a solution of 50 ml at a concentration of 500 cells/ml is obtained. An undiluted portion of this solution should be counted with an electronic counter to verify the cell number.

(iv) Further dilute the cells, using the direct count obtained in (iii) as a basis, so as to obtain one cell/4 ml. Add 1 ml of the diluted cells to each well of several 24-well culture dishes and incubate at 37°C for 7 days.

(v) With an inverted microscope check the number of colonies per well. If over 33% of the wells have cells, the process should be repeated because of the danger of multiple clones. If an appropriate number of colonies per well is found, screen each well for a single colony that has an epithelial morphology in crowded portions of the dish. Pick several candidates, remove by trypsin–EDTA treatment and expand into larger plates.

3. SPECIALIZED USES OF BALB/c-3T3 CELLS

3.1 **Mitogenesis**

BALB/c-3T3 cells are a valuable model for the investigation of cellular and molecular events that regulate cell cycle traverse. The technical ease with which the cells can be manipulated, the ability to obtain large numbers of cells synchronized at specific portions of the cell cycle, and the rigorous growth factor requirements are all attributes that make 3T3 cells particularly appropriate for these investigations.

The specific mitogens that stimulate proliferation in BALB/c-3T3 cells are outlined below. Although detailed procedures used to quantitate cellular proliferation are presented in another chapter, somewhat unexpected aspects of an assay used with 3T3 cells are also given.

3.1.1 *Growth factor requirements for BALB/c-3T3 cells*

The growth of BALB/c-3T3 is regulated by the synergistic action of distinct classes of mitogens. Platelet-derived growth factor (PDGF) is clearly the major serum-derived mitogen for mesenchymal cells (3). Treatment with PDGF alone, however, does not stimulate G_0/G_1 traverse in BALB/c-3T3 cells. The presence of either plasma-derived serum (4, 5) or a combination of EGF and insulin-like growth factor I (IGF-I) (6, 7) is also required for optimal growth. The phorbol ester 12-*O*-tetradecanoylphorbol-13-acetate (TPA) can replace the mitogenic activity of both PDGF and EGF and stimulate 3T3 cells to divide in the presence of only insulin-like activity (8, 9). BALB/c-3T3 cells, unlike 10T1/2, NRK or

AKR-2B cells, do not respond mitogenically to only EGF and IGF-I (10), in that they require prior exposure to PDGF for optimal growth. Even in cells responsive to EGF alone, however, a fundamental action of PDGF is to regulate the ability of the cells to respond to EGF (11).

Increased concentrations of intracellular cyclic AMP regulate the sensitivity of BALB/c-3T3 cells to EGF (10) in a manner similar to that observed in Swiss 3T3 cells (12). Cyclic AMP has also been shown to modulate responsiveness to PDGF-induced mitogenesis (13).

3.1.2 *Quantitation of proliferation in BALB/c-3T3 cells*

There are several methods suited for the measurement of proliferation in BALB/c-3T3 cells as outlined elsewhere in this book. In experiments in which arrested-cultures are stimulated to re-enter the cell cycle, the most informative data are the numbers of cells that divide. This can be precisely measured by autoradiography, flow microfluorimetry or cell counts. These direct measurements are often approximated by determining the amount of radiolabelled thymidine incorporated into a high-molecular-weight form during a short pulse. This technique is appropriate for measuring mitogenesis in BALB/c-3T3 cells as well, but some caution should be exercised when interpreting the data.

When density-arrested cells are stimulated to divide by any of a number of mitogens, the amount of [^{3}H]thymidine incorporated into trichloroacetic acid (TCA)-precipitable material during a 60-min pulse begins to increase at 12 h and is maximal at 24 h. When the percentage of S-phase cells determined by flow cytometric procedures is directly compared to the rate of thymidine incorporation at 24 h, the general pattern of the results is shown in *Figure 2*. There is a high rate of increase in thymidine incorporation when there are relatively few cells stimulated to divide. At higher rates of absolute cell growth, the relative increase in thymidine incorporation decreases. Therefore, an increase in thymidine incorporation consistently reflects an increase in the growth fraction, but it is important to note that the relationship between these two measurements is not linear.

3.2 **Quantitation of transformation**

One of the major uses of BALB/c-3T3 cells is in the investigation of the ability of any of a multitude of potential carcinogens, to induce all or part of the transformed phenotype. The exact characteristics of transformation can be somewhat elusive, and depend to some degree on the biases of the investigator. However, there is general agreement that the loss of anchorage-dependence and the abrogation of specific growth factor requirements are hallmark characteristics of transformation. Details of the procedures to measure these processes are outlined below.

3.2.1 *Growth in soft agar*

The ability of cells to grow in an anchorage-independent fashion is considered to be the classic predictor of tumorigenicity (14). The following is a simple procedure to directly measure anchorage-independence:

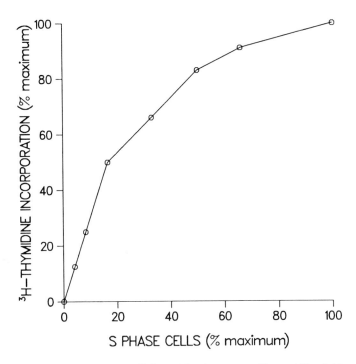

Figure 2. Comparison of the amount of [³H]thymidine incorporated into acid insoluble material and the number of cells in S-phase determined by flow microfluorimetry.

- (i) Add 1.4 g of agar to 100 ml of double-distilled water. The agar will not dissolve but it should be evenly suspended. Sterilize by autoclaving for 20 min followed by slow exhaust. Store at 4°C.
- (ii) Place the bottle of agar in a boiling water bath until the agar liquifies. Transfer the bottle to a 45°C water bath and allow it to equilibrate for at least 30 min.
- (iii) Add FCS to 2× αMEM to a final concentration of 20%. Prewarm to 37°C.
- (iv) Mix equal volumes of agar and 2× αMEM (final concentration 0.7% agar, 1× αMEM, 10% FCS), and pipette 1 ml into the bottom of a 35-mm Petri dish. Place in an incubator overnight.
- (v) Reliquify stock agar and equilibrate at 45°C as in (i) and (ii).
- (vi) Remove cells from the flask by treatment with trypsin–EDTA as described in Section 2.2.1. Resuspend in serum-free αMEM and remove an aliquot for a determination of cell number. Dilute the cell slurry with 37°C serum-free αMEM to a concentration such that 0.75 ml contains the number of cells to be plated in each dish.
- (vii) Mix one part agar, one part 2× αMEM containing 20% serum, and two parts cell slurry. The final concentration of agar in this solution is 0.35%. Quickly plate 1.5 ml over the 0.7% agar underlay. Gently rock the plate back and forth to evenly distribute the fresh agar.

(viii) After approximately 6 h, gently add 1 ml of αMEM supplemented with 10% serum over the agar layer.

 (ix) After 10–14 days the number of colonies containing more than 50 cells can be quantified with a low power dissecting microscope.

3.2.2 *Growth in methocel*

Although growth in agar is quite straightforward and easy to quantitate, some problems can arise. Like most mouse lines, BALB/c-3T3 cells are very sensitive to heat and cannot survive temperatures of over 39°C for even fairly short periods of time. To avoid potential problems with variable cell losses when 37°C cell slurries are mixed with 45°C agar solution, some laboratories utilize methocel rather than agar as a base. Medium containing this compound is a liquid at 4°C, and hardens as the temperature is raised to 37°C.

 (i) Add 4 g methocel to a 250-ml bottle. Put a stirring bar in the bottom and autoclave for 20 min followed by slow exhaust.

 (ii) Add 180 ml of medium that has been prewarmed to 56°C.

 (iii) Stir at room temperature for 30 min, and at 4°C overnight or until clear.

 (iv) Add 20 ml of FCS which brings the final concentration to 10%.

 (v) Centrifuge in a T160 rotor at 15000 r.p.m. for 30 min.

 (vi) Pour off the supernatant and store at 4°C.

 (vii) Prepare 35-mm Petri dishes with a 0.7% agar underlay as described in Section 3.2.1 through step (v).

(viii) Remove cells from plates with trypsin–EDTA as described in Section 2.2.1, resuspend in αMEM plus 10% FCS, and remove an aliquot to determine the cell number. Dilute the cells at 4°C with medium supplemented with 10% serum such that the number of cells to be plated in each dish is contained in 0.5 ml.

 (ix) Add three parts chilled methocel to one part cell slurry and gently mix to obtain an even distribution of cells. Add 2 ml of the slurry to each 35-mm plate and immediately place in a 37°C incubator.

 (x) After 10–14 days quantitate the number of colonies containing more than 50 cells.

3.2.3 *Growth for cells in plasma-derived serum*

Non-transformed cells require multiple serum-derived growth factors to grow at optimal rates (15). Although transformation can potentially abrogate the need for any of several of these mitogens, typically transformed BALB/c-3T3 cells have a rather specific loss of their requirement for PDGF (16). The alteration can easily and specifically be measured by growth in plasma-derived serum (PDS), some-times referred to as platelet-poor plasma (PPP). PDS can be used as a medium supplement at a concentration ranging from 3 to 10%. Untransformed BALB/c-3T3 cells will double less than twice over a 7-day period in PDS-supplemented medium. By contrast, several classes of transformed 3T3 cells will grow at equivalent rates in serum and plasma. Cells sparsely plated in medium containing whole blood serum, however, will not become growth-arrested when fresh medium containing PDS is added. When cells are not confluent, sufficient quantities of PDGF will bind to uncovered plastic such that significant growth will be obtained even in the subsequent presence of PDS. There are now commercial sources of both equine and bovine PDS, but experiments can easily be performed using locally produced human material.

(i) Obtain 50-ml samples of freshly drawn venous blood. The blood should be drawn using an 18-gauge butterfly needle and a 50-ml syringe. Do not use a Vacutainer since is causes a significant platelet lysis and the subsequent release of PDGF. Place the blood in 50-ml conical tubes and immediately submerge it in crushed ice for approximately 20 min. If even small amounts of clot formation are observed, the sample should be discarded or used to prepare whole blood serum.

(ii) Centrifuge the blood at 4°C at 2400 r.p.m. for 20 min in a clinical centrifuge. Remove the clear plasma fraction, being careful not to obtain any of the buffy coat. This is most easily achieved if 2–3 ml of plasma are left above the packed red cells.

(iii) Aliquot the plasma into 15-ml sterile conical tubes, break off about 2.5 cm of the end of a sterile Pasteur pipette into each tube, and incubate at 37°C for approximately 2 h. A solid clot should form.

(iv) Insert a sterile 9 inch Pasteur pipette approximately 0.5 inches into the plasma clot and begin to swirl it along the outer rim of the tube in a circular motion. The clot will begin to adhere to the pipette and, as it does, lower it further into the tube. Eventually a pea-sized fibrous clot, containing the broken tip, will remain at the bottom of the tube.

(v) The plasma should be sterile-filtered, heat inactivated at 56°C for 30 min, and stored at −20°C.

3.3 Tumorigenicity

The exact relationship between *in vitro* transformation properties and *in vivo* tumorigenicity is still not well defined at either a descriptive or a functional level. The use of *in vitro* phenotypes such as growth in soft agar or the abrogation of specific growth factor requirements can provide a wide spectrum of information concerning alterations caused by specific molecular insults. However, the further development of several important concepts still requires whole animal tumorigenicity studies. Areas such as angiogenesis, tumour progression or host cell immune responses can be thoroughly investigated only in intact animals.

3.3.1 *Conventional tumorigenicity assay*

Fairly straightforward questions concerning the tumorigenicity of 3T3 cells expressing various degrees of the transformed phenotype can be asked with a standard subcutaneous injection system.

(i) Remove the cells to be injected from the culture plates with trypsin–EDTA as described in Section 2.1.1.

(ii) Add 5 ml of medium containing 10% FCS to each plate and pipette the slurry up and down to disrupt any clumps. Pool the medium from plates containing identical cells and remove an aliquot for a determination of the cell number.

(iii) Centrifuge the cell slurry for 5 min at 1000 r.p.m. in a clinical centrifuge and resuspend in serum-free medium in volume such that the number of cells to be injected is present in 0.1 ml.

(iv) Inject the cells subcutaneously either into a syngeneic BALB/c mouse or a nude mouse. There are several suitable injection sites, but the flank is technically easy and a subsequent tumour at that site is easily detectable at an early stage. Depending on the cell type and the expected tumorigenicity, as few as 10^3 cells can be injected. A cell should not, however, be designated to be non-tumorigenic unless 10^6 cells are injected into each of five animals which are then observed to be tumour-free for at least 90 days.

3.3.2 *Tumorigenicity using preimplanted sponges*

Tumour progression, defined as long latency tumorigenicity with a subsequent altered phentoype acquired *in vitro*, is difficult to study using the subcutaneous injection assay described above. Cells injected without a support typically form tumours rather rapidly or not at all. The injection of cells into preimplanted Gelfoam sponges (17) can provide information concerning *in vivo* progression since with this method tumours can be observed as late as a year following the injection of cells. It has been documented that long latency tumours can be found in animals using a sponge assay with cells that are not tumorigenic by conventional assays (17).

(i) Cut sterile Gelfoam (Upjohn Co.) sponges with a scalpel into $16 \times 17 \times 10$ mm segments and hydrate with serum-free medium for 20 min at room temperature.

(ii) Anaesthetize mice with pentabarbitol (60 mg/kg) injected i.p. Make a 1-cm incision dorsally across the animal's back and insert a prehydrated sponge subcutaneously. Close the wound with a 9-mm autoclip.

(iii) Following a 10-day period, in which the sponge becomes vascularized, prepare cells as described above and inject a 0.1-ml aliquot as close to the centre of the sponge as possible.

(iv) Animals used in sponge assays should be observed for tumours for a minimum of 9 months.

4. ACKNOWLEDGEMENTS

We would like to thank Patrick O'Shea, Paul Kraemer, Judy Tesmer, Elizabeth Saunders, Janet Cooper, Robert Tobey, Evelyn Campbell and Brad Stone for help during the preparation of this manuscript.

5. REFERENCES

1. Todaro, G. and Green, H. (1963) *J. Cell Biol.*, **17**, 299.
2. House, W. and Waddell, A. (1967) *J. Path. Bact.*, **93**, 125.
3. Ross, R. and Vogel, A. (1978) *Cell*, **14**, 203.
4. Pledger, W. J., Stiles, C. D., Antoniades, H. N. and Scher, C. D. (1977) *Proc. Natl. Acad. Sci. USA*, **74**, 4481.
5. Vogel, A., Raines, E., Kariya, B., Rivest, M. J. and Ross, R. (1978) *Proc. Natl. Acad. Sci. USA*, **75**, 2810.
6. Leof, E. B., Wharton, W., Van Wyk, J. J. and Pledger, W. J. (1982) *Exp. Cell Res.*, **141**, 107.
7. Leof, E. B., Van Wyk, J. J., O'Keefe, E. J. and Pledger, W. J. (1983) *Exp. Cell Res.*, **147**, 202.
8. Frantz, C. N., Stiles, C. D. and Scher, C. D. (1979) *J. Cell. Physiol.*, **100**, 413.
9. Selinfreund, R. and Wharton, W. (1986) *Cancer Res.*, **46**, 4486.
10. Olashaw, N. E., Leof, E. B., O'Keefe, E. J. and Pledger, W. J. (1984) *J. Cell. Physio.*, **118**, 291.
11. Wharton, W., Leof, E. B., O'Keefe, E. J. and Pledger, W. J. (1982) *Exp. Cell Res.*, **147**, 443.
12. Rozengurt, E. (1982) *J. Cell. Physiol.*, **112**, 243.
13. Wharton, W., Leof, E. B., Olashaw, N., Earp, H. S. and Pledger, W. J. (1982) *J. Cell. Physiol.*, **111**, 201.
14. Freedman, V. H. and Shin, S. (1974) *Cell*, **3**, 355.
15. Scher, C. D., Shepard, R. C., Antoniades, H. N. and Stiles, C. D. (1979) *Biochim. Biophys. Acta*, **560**, 217.
16. Pledger, W. J. (1984) In *Control of Cell Growth and Proliferation*. Veneziale, C. M. (ed.), Van Nostrand Reinhold, New York, p. 108.
17. Wells, R. S., Campbell, E. W., Swartzendruber, D. E., Holland, L. M. and Kraemer, P. M. (1982) *J. Natl. Cancer Inst.*, **69**, 415.

Suppliers of commonly used cell lines

Cell line	Origin	Remarks
L-M (TK⁻)	Connective tissue, mouse	Thymidine kinase deficient
HeLa	Epithelial carcinoma, human	Suspension and monolayer cultures
CCRF-S-180II	Sarcoma 180, mouse	Chemotherapy screening tests
BHK-21	Kidney, Syrian hamster	Fibroblast-like, contact-inhibited
Don	Lung, diploid, Chinese hamster	Fibroblast-like
KB	Epidermoid carcinoma, human	Like HeLa
BS-C1	Kidney, African green monkey	Epithelial-like
MDCK	Kidney, canine	Epithelial-like, viral studies
Detroit 532	Skin, human fibroblasts	Down's syndrome, male
Pt K2	Kidney, marsupial	Only 13 chromosomes
CV-1	Kidney, African green monkey	For SV40 growth
WI-38	Lung diploid fibroblasts, human	Finite life-span
2RA	Lung, human	SV40-transformed WI-38
GH₃	Pituitary tumour, rat	Somatotrophin and prolactin-secreting
EB-3	Burkitt lymphoma, human	Suspension
Raji	Burkitt lymphoma, human	Suspension
3T3-Swiss	Embryo, mouse	Contact-inhibited fibroblasts
3T6-Swiss	Embryo, mouse	Collagen-secreting
C6	Glial tumour, rat	S-100 protein-secreting
HT-1080	Fibrosarcoma, human	Transformed
BALB/c3T3 c1A31	Embryo, mouse	Contact-inhibited non-tumorigenic
MRC-5	Lung, diploid human fibroblasts	Like WI-38
IMR-90	Lung, diploid human fibroblasts	Like WI-38
Daudi	Burkitt lymphoma, human	Suspension
Colo 320 DM	Colon, adenocarcinoma, human	Tumorigenic
C3H 10T1/2, clone 8	Embryo, mouse	Contact-sensitive fibroblasts
HL-60	Human, promyelocytic leukaemia	Suspension
K-562	Human, chronic myelogenous leukaemia	Suspension
KG-1a	Human, acute myelogenous leukaemia	Suspension

In addition, human fibroblasts from a great variety of genetic disorders are available from Institute for Medical Research, Copewood and Davis Streets, Camden, NJ 08103, USA

Index